Challenges in Physics Education

Series Editor

Marisa Michelini, Dipartimento di Fisica, University of Udine, Udine, Italy

This book series covers various aspects of physics teaching and learning across all educational levels and learning environments. The individual volumes cover a wide array of topics, including innovative approaches and pedagogical strategies for physics education, the development of effective methods to integrate multimedia into physics teaching and learning, pioneering lab experiments, and the utilization of web-based interactive activities. Selected research-based contributions will be prominently featured throughout.

The GIREP, the International Research Group on Physics Teaching, collaborates on this publication series, incorporating selected papers from globally recognized experts and monographs. Book proposals from other sources are also encouraged.

"Challenges in Physics Education" is targeted towards professionals, teachers, researchers, instructors, and curriculum developers, aiming to enhance physics teaching and learning and, consequently, the overall perception of physics within society.

Proposals for publication should be submitted to Marina Forlizzi (marina.forlizzi@springer.com).

Italo Testa · Marisa Michelini · Salvatore Esposito
Editors

Connecting Physics Education Research and Practice

Springer

Editors
Italo Testa
Department of Physics "Ettore Pancini"
Federico II University of Naples
Naples, Italy

Marisa Michelini
University of Udine
Udine, Italy

Salvatore Esposito
Department of Physics "Ettore Pancini"
Federico II University of Naples
Naples, Italy

ISSN 2662-8422 ISSN 2662-8430 (electronic)
Challenges in Physics Education
ISBN 978-3-031-86608-1 ISBN 978-3-031-86609-8 (eBook)
https://doi.org/10.1007/978-3-031-86609-8

This Springer imprint is published by the registered company Springer Nature Switzerland AG
The registered company address is: Gewerbestrasse 11, 6330 Cham, Switzerland

Committees

Editors

Italo Testa
Marisa Michelini
Salvatore Esposito

Scientific Editorial Board

Claudio Fazio, University of Palermo
Marco Giliberti, University of Milan
Marisa Michelini, University of Udine
Stefano Oss, University of Trento
Nadia Robotti, University of Genova and Centro Fermi Roma
Italo Testa, University of Neaples Federico II

External Scientific Reviewers

Micol Alemani (Germany)
Marina Garzon Barrios (Colombia)
Philipp Bitzenbauer (Germany)
Joan Borg Marks (Malta)
Digna Couso (Spain)
Mieke De Cock (Belgium)
José Cantó Doménech (Spain)
Pandora Dorouka (Greece)
Sergej Faletic (Slovenia)

Reguiero Bibiana Fernandez (Spain)
Jenaro Guisasola (Spain)
Meli Kalliopi (Greece)
Grzegorz Karwasz (Poland)
Zuzana Jeskova (Slovakia)
Eilish McLoughlin (Ireland)
Martín Monteiro (Uruguay)
Maja Planinic (Slovenia)
Bianca Reinisch (Germany)
Dagmara Sokolowska (Poland)
Fatih Tasar (Türkiye)
Edit Yerushalmi (Israel)
Kristina Zuza (Spain)

Preface

This book presents a selection of the most recent research results of the CooFIS08 community, which represents Italian university colleagues working in the fields of physics education and the history of physics who have decided to work in a coordinated way. Many of them have been present in the international literature on physics education research for years, even in positions of responsibility. The motivation for this publication stems from the desire to offer an operational contribution to school practice to the world of schools and researchers in scientific disciplines that contribute to the improvement of teaching at a pedagogical, psychological or sociological level. There is a great need for reference materials for teachers on the results of research in science education. In fact, the main commitment of researchers is usually to offer their results in a non-operational and specialized way, and in specialized journals, which are often not accessible or usable by teachers. It is precisely this operational role that often bridges the gap between research and teaching and, when translated into conceptual terms, constitutes a training for teachers, guiding them towards ways of improving the learning contexts and processes to be put in place in physics education in order to promote learning. The selection of those researches in physics education and history that make the link between research and practice also means dialogue with other research areas that deal with teaching, and a mutual fertilization of research.

The research results collected here were presented at the second meeting of CooFIS08—Italian Physics Education Research and History of Physics Research (IPER-HPR), organised in Naples by CooFIS08, University of Naples Federico II (30 November—01 December 2008) in Naples and sponsored by the Italian Physical Society. The Local Organising Committee received the submissions, which were subjected to the Editorial Board and then to a rigorous process of double-blind peer review by international referees to select the papers presented here.

The first part of the book includes two contributions on teaching physics and science at the primary school level, with a focus on gender-related aspects. The first contribution investigates children's gendered perceptions of scientists at work, addressing aspects of the Nature of Science. The second focuses on teaching energy concepts, emphasizing children's self-efficacy. Understanding children's gendered perceptions of scientists and enhancing their self-efficacy in learning energy concepts are crucial for fostering early interest and confidence in science. The second part underscores the importance of investigating students' ideas when learning physics at the secondary school level. By examining students' reasoning about electromagnetic waves and their difficulties with mathematical concepts like derivatives, integrals and vectors, educators can develop more effective teaching strategies to address these challenges. The third part emphasizes the development of teaching-learning modules for advanced physics topics, such as surface phenomena and the greenhouse effect. These topics are rarely covered in the secondary school curriculum, making their inclusion vital for providing students with a comprehensive understanding of physics. The fourth part focuses on the use of laboratories in teaching physics, highlighting the importance of hands-on learning. The contributions on using smartphones, augmented reality devices, Python software, and real-time devices, as well as exploring geometric properties of light reflection, demonstrate innovative approaches for engaging students in practical experiments. The fifth part stresses the value of incorporating the history of physics into teaching modern physics. By examining the role of old quantum mechanics and the steps Planck took to establish his renowned constant, students gain a deeper appreciation of the historical context and development of key scientific concepts. The sixth part discusses two research studies on student engagement in learning physics. The first contribution examines how academic engagement affects students' performance in their first year of university. The second one addresses the transition from high school to university, focusing on developing methodological tools to support student retention in undergraduate STEM courses. These studies provide insights into how supporting students' performance and retention in undergraduate STEM courses, which is critical for their long-term success. Finally, the seventh part features four contributions on teacher education. The first three papers address proposals for teaching force, energy, and properties of matter at the primary school level, while the fourth one proposes a training course for in-service lower secondary school teachers in classical mechanics. The contributions about how teaching force, energy, and properties of matter at the primary school level, as well as the training course for in-service lower secondary school teachers in classical mechanics, underscore the need for continuous professional development to enhance teaching effectiveness.

We hope that this book will be a valuable resource for teachers, teacher educators, researchers and policy makers working to improve physical education in many countries.

Naples, Italy — Italo Testa
Udine, Italy — Marisa Michelini
Naples, Italy — Salvatore Esposito

Contents

Perception and Science Teaching in Primary School

From DAST to DPFS: An Overview and Preliminary Results of a National Survey on the Perception of Science in Primary School

Peppino Sapia, Onofrio Rosario Battaglia, Claudio Fazio, and Giacomo Bozzo

Abstract Public perception of science plays a crucial role in shaping societal attitudes towards scientific advancements and technologies. This perception, particularly among young children, significantly influences their learning experiences in scientific disciplines, impacting their career aspirations and choices of higher education.

In the 1980s, a noteworthy survey known as the "Draw A Scientist Test" (DAST) was conducted across Canada, USA, and Australia with the aim of profiling primary school children's perceptions of science and scientists. Recently, the DAST and its subsequent iterations have regained attention in international educational research. This increasing interest is concurrent with a concerning trend of declining interest in scientific careers observed in many countries all around the world.

In response to these trends, the laboratory for Science Communication "AgoràLAB", in collaboration with the inter-university research centre "GEO" started the study "Draw a Person Doing Science" (DPFS) in Italy. This study involved a sample of 1732 primary school children (50% male, 50% female; geographical distribution: 37% northern Italy, 9% centre, 54% southern), who were tasked with drawing persons engaged in scientific activities in their workplaces.

This paper presents the national project, describes a preliminary analysis of a subset ($N = 1101$) of the sample and discusses the first findings obtained by analysing the drawings on the base of over 20 predefined indicators. The analysis confirms some stereotyped representations discussed in literature (the scientist is mainly represented as a male person working in an indoor laboratory) and features new characteristics (scientists are middle aged, with a neat hairstyle).

P. Sapia · G. Bozzo (✉)
AgoràLAB: Laboratory for Science Communication, University of Calabria, Arcavacata, Italy
e-mail: giacomo.bozzo@unical.it

Department of Mathematics and Computer Science, University of Calabria, Arcavacata, Italy

O. R. Battaglia · C. Fazio
Department of Physics and Chemistry - Emilio Segrè, University of Palermo, Palermo, Italy

I. Testa et al. (eds.), *Connecting Physics Education Research and Practice*,
Challenges in Physics Education, https://doi.org/10.1007/978-3-031-86609-8_1

1 Introduction

In an era where technological progresses shape our daily life, scientific literacy (SL) is crucial for fostering a society that has to deal with complex global challenges and represents a first important step of lifelong learning process. Being scientifically literate offers everyone the opportunity to evaluate the implications of scientific advancements on their daily life, society, and on the environment, and is crucial for promoting a more responsible citizenship and a sustainable future. SL involves not only the ability to understand scientific concepts but also to critically evaluate information, make informed decisions, and participate appropriately in discussions (Miller 2012, 2016; Sapia 2019; Turiman et al. 2012). In this perspective, it is more appropriate to introduce the idea of Civic Scientific Literacy (CSL) (Sapia 2019), which represents "a cornerstone of informed public policy" (Shen 1975).

Fostering SL in primary schools represents the ground steps for a future generation that is not only familiar with scientific principles but is also equipped to contribute meaningfully to the advancements and challenges of our dynamic and scientifically-driven society. SL in primary education goes beyond memorizing facts; it encourages students to explore, ask questions, and engage in hands-on experiments. This early exposure not only increases interest in various scientific disciplines but also fosters the skills necessary to analyse and interpret the ever-expanding realm of scientific knowledge.

However, the perception of science (even in childhood) influences the starting steps of SL. When children view science as exciting and accessible, they are more likely to develop a positive attitude towards learning scientific concepts. Conversely, negative perceptions about science can act as obstacles to the development of SL.

2 History of the Draw a Scientist Test

Different studies have investigated children's perception of science, with the aim of understanding how their perception influence their scientific learning. The analysis of stereotypes assigned to scientists has been conducted since 1905s. In a pioneering study, Mead and Métraux showed the typical perception of high school students about scientists, described as elderly males wearing eyeglasses and a white lab coat, who perform dangerous experiments in a laboratory (Mead and Métraux 1957). Several studies have been conducted across the time to analyse students' perception and various assessment tools have been planned and tested. One of the most used tests about the perception of science is the "Draw A Scientist Test" (DAST), developed in the 1980s by Chambers (1983), which has been used in many studies to investigate and analyse preconceptions and stereotypes associated with the image of a scientist in pupils' minds (Bozzato et al. 2021; Farland-Smith 2009, 2012; Ferguson and Lezotte 2020; Finson 2003, 2012; Finson et al. 1995; Miller et al. 2018; Reinisch et al. 2017;

Symington and Spurling 1990). This test was introduced to investigate how cultural and social stereotypes could influence expectations and views of scientific careers.

The DAST was developed with the aim of evaluating children's conceptions of scientists through the analysis of the drawings they make on the basis of an appropriate instruction. Usually, children find it more natural to express their thoughts, feelings, and ideas through drawing rather than through writing or other forms of verbal communication (especially in the first years of primary school). In fact, as elsewhere highlighted (Fabris et al. 2023) "it is believed that drawing can be a window into the child's inner world, allowing to capture feelings, representations, and perceptions related to a specific topic of investigation".

The analysis of the previously discussed literature offered the opportunity to identify stereotyped images of scientists (characterized by 7 indicators present in pupils' drawing: Lab coat, Eyeglasses, Facial growth of hair, Symbols of research, Symbols of knowledge, Technology, Relevant captions) and when such images began to appear in pupils' ideas.

For several decades, the DAST has been widely used and different studies have been conducted with the aim of analysing a series of methodological variations.

Symington and Spurling (1990) changed the instruction "Draw a Scientist" in "Do a drawing that tells what you know about scientists and their work", with the aim of studying the idea of pupils about the job of scientists instead of focusing on scientists themselves (Symington and Spurling 1990).

Several studies tried to improve the objectivity and reliability of the DAST proposed by Chambers. One of them was conducted by Finson et al. (1995), in which they proposed a checklist for evaluating the DAST drawing: the Draw-A-Scientist Checklist (DAST-C) (Finson et al. 1995). In this work, they added 7 indicators to those introduced by chambers (i.e. Gender, Race, Indicators of Danger, Presence of Light Bulbs, Mythic Stereotypes, Indicators of Secrecy, Indoor or Outdoor work).

A study conducted in 2012 (Farland-Smith 2012) furtherly changed the instruction "Draw a Scientist" in: "Imagine that tomorrow you are going on a trip (anywhere) to visit a scientist in a place where the scientist is working right now. Draw the scientist busy with the work this scientist does. Add a caption, which tells what this scientist might be saying to you about the work you are watching the scientist do. Do not draw yourself or your teacher".

The modified DAST (i.e. mDAST) demonstrated that, by detailing the instruction and by adding specific questions (the gender of the children, the gender of the scientist, the place of the scientist, the job of the scientist), it is possible to stimulate pupils towards more detailed mental images of scientists.

Many studies have validated the DAST-C tool in different contexts. An interesting analysis about the use of DAST-C was conducted by Ferguson and Lezotte (2020). They demonstrated that DAST-C generally appears appropriate to measure stereotypes and students' perceptions of scientists (Ferguson and Lezotte 2020). However, they suggested some revisions to the DAST-C, in terms of (a) clearer instructions and

(b) additional categories. In particular, analysing the literature, they identified 58 individual categories collapsed into four general themes: "Anti-Social/Eccentric", "Passionate/Happy", "Curious/Hard Working", and "Intelligent" (Ferguson and Lezotte 2020).

3 Gender Issues and a New Version of DAST: DPFS

International research literature about DAST has provided valuable insights into the evolution over time of students' perceptions about scientists. In addition, many studies have also put in evidence an important gender issue, that reflects broader challenges faced by women in STEM fields, starting as early as primary school, where gender biases and societal expectations can significantly influence career aspirations. These issues have been pointed out by several studies demonstrating the persistence of gender stereotypes that influence children's perceptions and expectations regarding STEM careers (Ferguson and Lezotte 2020; Losh et al. 2008). It is crucial to address these gender stereotypes from the early years of primary school, promoting an educational environment that challenges traditional gender-based roles and encourages all students to explore STEM disciplines without gender-based limitations. In fact, pupils also in primary schools (both males and females) "are more likely to link scientific disciplines to males and believe that males are better at science" (Ferguson and Lezotte 2020; Lane et al. 2012; Miller et al. 2015). In addition, STEM-related fields are perceived as male-dominated (Ferguson and Lezotte 2020; Hill et al. 2010) and characterized by unattractive lifestyles (Ferguson and Lezotte 2020; Miller et al. 2006), which contribute to the STEM gender gap (Ferguson and Lezotte 2020). Another interesting finding is represented by the fact that the gender bias in the perception of science increases as children grow older, even if women's representation in science has increased in the last decade (Miller et al. 2018).

The research described in this paper was designed and carried out within the "AgoràLAB" project at the University of Calabria,[1] in collaboration with the GEO inter-university research centre,[2] with the aim of investigating the perception of science and the possible persistence of gender stereotypes in modern society, despite various awareness campaigns have been conducted in the last decades by various political and cultural institutions. Starting from the analysis of the categories of the DAST and its evolutions, our work was planned (a) to verify which categories may still be useful and which are obsolete and could be discarded and (b) to analyse the problems related to gender stereotypes in Italian schools. In this perspective, we

[1] "AgoràLAB: Laboratory for the dissemination of technical-scientific culture for the XXI century citizenship" is an initiative funded by the Italian Ministry for University and Research (MUR) under grant n. ACPR20_00229.

[2] GEO is the "Italian Interuniversity Research Centre for the Study of Youth Condition, Educational Institutions and Guidance". Website: https://geo.uniud.it/.

reformulated the original instruction "Draw A scientist", because its translation in Italian introduces to gender issue, considering that the term "scientist" in Italian is usually translated as "Scienziato", which represents a male person. To resolve this problem, the instruction was rephrased in Italian as "Disegna una Persona che Fa Scienza" (DPFS) literally meaning "Draw a Person Doing Science". More in details, starting from the study proposed in 2012 (Farland-Smith 2012), we proposed to primary schools' students the following instruction: "*Imagine going on a school trip to visit a scientist at his/her workplace and draw him/her doing his/her work. Add a comic to the drawing, or a separate caption, in which the scientist you have drawn talks about the work he/she is doing. Don't draw yourself or your teacher*". The main goal of such a rephrasing was to avoid any kind of influence regarding the gender of the scientist to be drawn.

In addition, each pupil enrolled in our study was provided with a form (Fig. 1) with the aim of collecting additional information regarding: *(i)* some features of the drawn scientist and *(ii)* some relevant information about pupils' parents. More in details, as regards the first point, we collected additional and explicit information on the drawn person, as for example the intended gender, the workplace (indoors or outdoors) and the activities done by the scientist. As regards the second point, some anonymous information about pupils' parents (i.e. the education levels and the employments) was collected by the primary school teacher.

4 Context and Methodology of the Study

The research presented in this paper had been planned by a national network of research groups in physics education with the aim of enrolling a wide sample of primary school students from north to south of Italy and of collecting (consequently) a sample of drawings representative of different geographical areas, historically having different and well-known socio-economic characteristics. Figure 2 shows the distribution of the cities involved in this project. The collected drawings (1732) represent a rich sample, which offers the opportunity to compare illustrations made by pupils of the three different parts of Italy, with the aim of analysing the impact of social and cultural elements in pupils' perceptions of science and of persons who practice it.

In this paper we present the results of the analysis of a partial subset of the whole sample, consisting in $N = 1101$ drawings. Figure 3 illustrates the geographical composition and the gender composition of the subset. More than 2/3 of the enrolled primary school students belongs from the schools of south of Italy (68%), equally representing the male (50%) and female gender (50%). Such equal gender representation cannot be found in the sample of students belonging from north (53% of male and 47% of female students) and centre of Italy (54% of male and 46% of female students).

The drawings were collected by the primary teachers of the schools involved in this project, who had been appropriately instructed before proposing the test to their students. All the material produced by pupils (from all the enrolled Italian

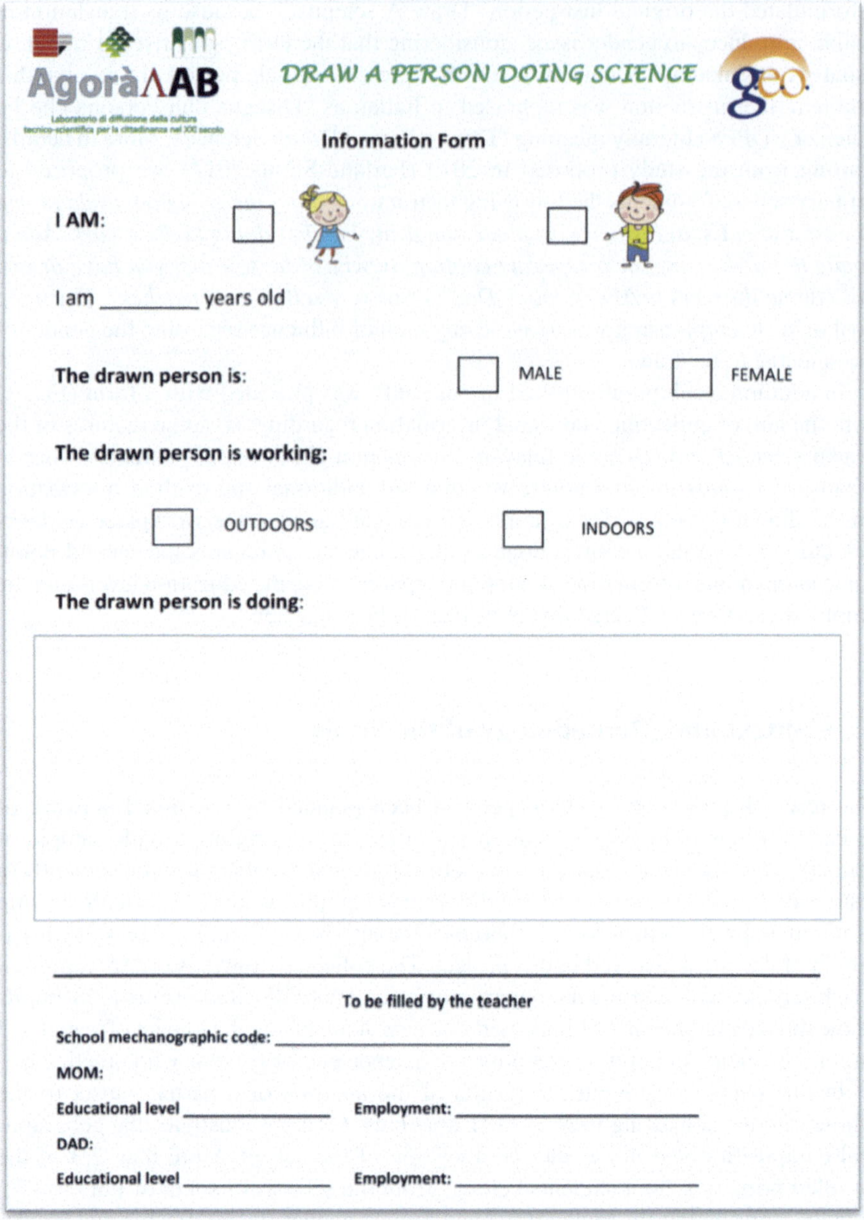

AgoràLAB
Laboratorio di diffusione della cultura tecnico-scientifica per la cittadinanza nel XXI secolo

DRAW A PERSON DOING SCIENCE

geo.

Information Form

I AM: ☐ ☐

I am ________ years old

The drawn person is: ☐ MALE ☐ FEMALE

The drawn person is working:

☐ OUTDOORS ☐ INDOORS

The drawn person is doing:

To be filled by the teacher

School mechanographic code: ____________________

MOM:

Educational level ____________ Employment: ____________________

DAD:

Educational level ____________ Employment: ____________________

Fig. 1 The figure shows the "Information Form": the first part of the Form offers the opportunity to collect additional information of the drawn person (the gender, the workplace, the activities that the scientist is doing); the second part permits to fill in some anonymous information about pupils' parents (the education levels and the employments)

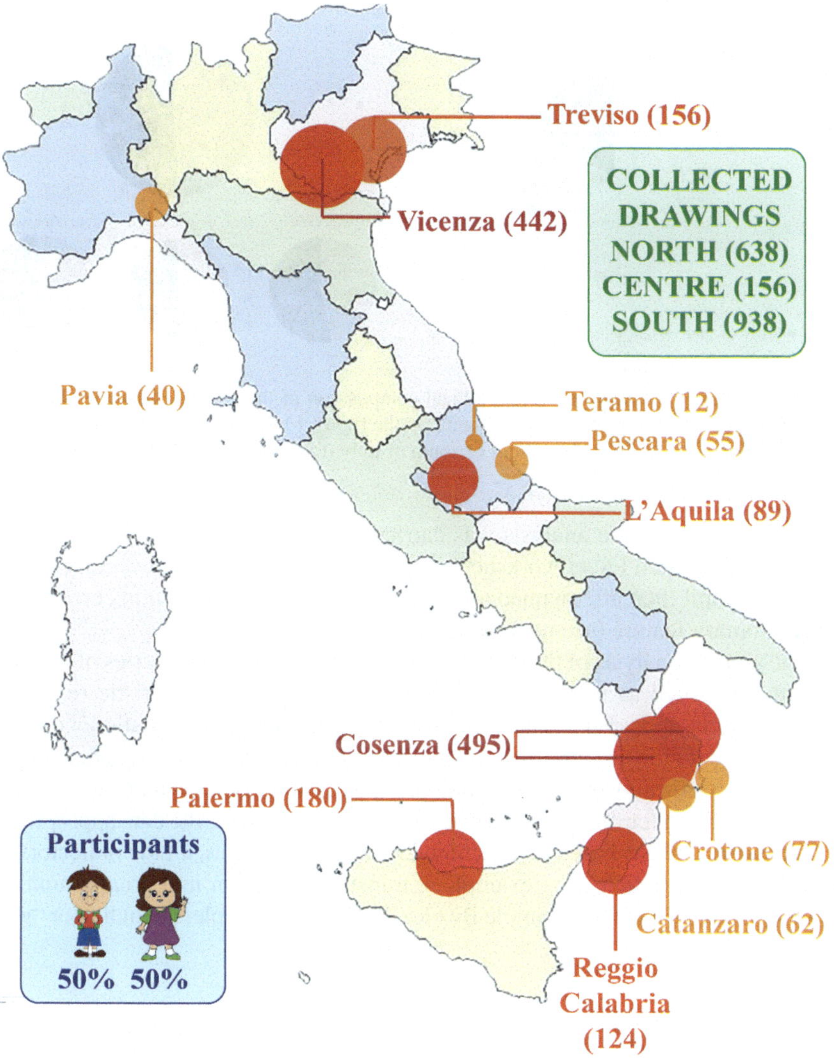

Fig. 2 The figure shows the distribution of the collected drawings made by primary school pupils (Participants) in different Italian cities

schools) was collected by the laboratory AgoràLAB at the University of Calabria. The analysis of such drawings was performed by a group of students of the last year of the master's degree course in Primary Education Sciences at the University of Calabria (15 students enrolled in different periods, in two different academics years, hereinafter referred to as ANALYSTS), who had been provided with a unified protocol of analysis.

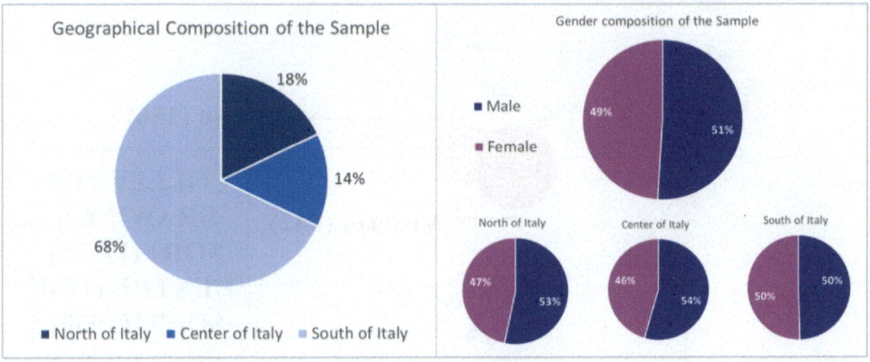

Fig. 3 The figure illustrates the geographical composition of the sample (figure on the left) and the gender composition of the sample (figure on the right). More than 2/3 of the enrolled primary school students belong from the school of south of Italy (68%)

More in details, the analysis was carried out by recording the features of each drawing in a shared Excel worksheet (Fig. 4), in which (a) each row corresponds to an author pupil (hereafter named a "datapoint") and (b) each column corresponds to a given binary feature (see next paragraph).

The graphic analysis of the drawings was carried out based on series of indicators, roughly classified as follows: *(a)* well-defined graphic characteristic (e.g. books, scientific tools, appearance of the scientist's hair, etc.), *(b)* characteristics of the author and of his/her family (gender, degree of instruction of parents, etc.), (c) characteristics of the work environment (indoor, outdoor), *(d)* personality traits that can be deduced from the drawing (curious, extravagant, messy, etc.). To make the data appropriate for a successive analysis based on Machine Learning techniques, all the indicators were binarized (see the following examples), transforming them in a greater number of binary "features" (see for example Bezdec 2021). For example, the indicator "gender

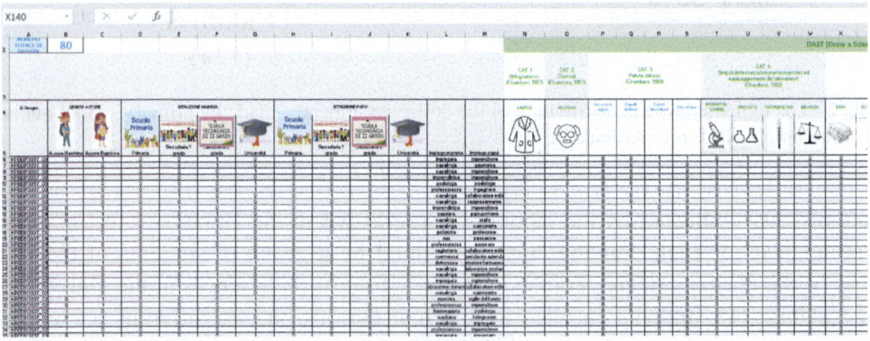

Fig. 4 The unified protocol of analysis was performed through a shared Excel worksheet, designed to analyse the presence/absence of the indicators used in our research. Each raw corresponds to a pupil's drawing (a datapoint) and each column corresponds to a binary feature of the drawing

of the author" is binarized by splitting it in the two features "author is male" and "author is female", each of which could only take on the value 1 or 0. Analogously, the indicator "Parents' level of education" is split into six binary features as: "mother, lower secondary school", "mother, upper secondary school", "mother, degree", and the corresponding three more feature for the father. Table 1 summarizes the taxonomy of the 87 features. This number does not mean that 87 characteristics of the drawings were considered, given the described binarization process leading from indicators to features.

A cross-analysis was planned and performed to verify the consistency of the analyses performed by the Analysts. To this end, the analysts were divided in two groups, respectively 10 students (Main team, MT) and 5 students (Crosscheck team, CT). Each analyst in MT was assigned approximately 170 drawings. A subset of 750 drawings, randomly picked out of the whole set analysed by the MT (1732), was randomly assigned to the CT (150 drawings pro-capite). In this way, 43% of the

Table 1 Taxonomy of the indicators used for the analysis (features) of the drawings

N. of the feature (f_n)	Macro-category (subset) of the features space	Examples	Conventional designation of the subset	Dimensionality of the subset
1–12	Personal, social, family characteristics of the author	Gender, age, parents	A	$D_A = 12$
13–34	Basic features of the drawn person and of the work environment (Chambers 1983)	Clothing, hair, symbols of research and knowledge, relevant writings and symbols, etc.	B	$D_B = 22$
35–49	Additional features of the drawn person and of the work environment. Stereotypes (Finson et al. 1995)	Gender and age, indication of danger or secrecy, mythic stereotypes, indoor/outdoor, …	C	$D_C = 15$
50–80	Personality traits of the drawn person (Ferguson and Lezotte 2020)	Smart, crazy, bored, curious, extravagant, messy, …	D	$D_D = 31$
81–87	References to specific scientific and technological disciplines (These authors 2024)	Physics, chemistry, mathematics, engineering, biology, medicine, natural sciences	E	$D_E = 7$

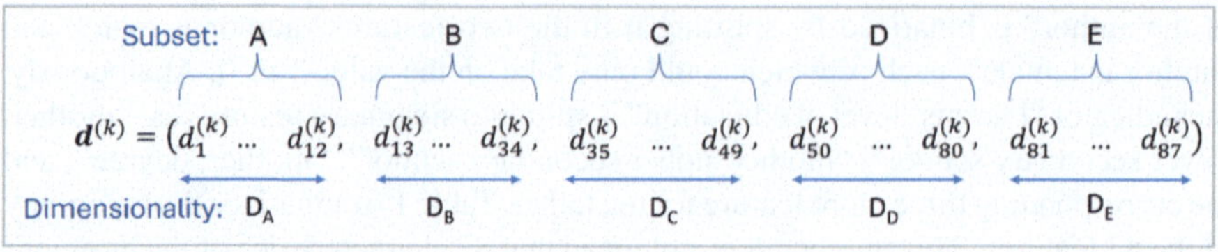

Fig. 5 The kth drawing of the dataset is encoded in an 87-components vector (a sequence of 0 and 1). Components can be grouped into the 5 subsets of features given in Table 1

sample was independently analysed by the two teams and the results was compared in order to assess the reliability of the analysis, which, in any case, may be affected by elements of subjectivity in the recognition and classification of graphic elements. The degree of consistency of the analysis was quantitatively evaluated as follows.

Given the data encoding above described, a datapoint of the dataset (i.e. a given drawing, for example the kth) corresponds to a vector having $N = 87$ components (Fig. 5).

In order to check the consistency of the analysis in each subset of the features space (see Table 1), the vector in Fig. 5 could be usefully rewritten as

$$\boldsymbol{d}^{(k)} = \left(\boldsymbol{d}_A^{(k)}, \boldsymbol{d}_B^{(k)}, \boldsymbol{d}_C^{(k)}, \boldsymbol{d}_D^{(k)}, \boldsymbol{d}_E^{(k)}\right) \tag{1}$$

where

$$\boldsymbol{d}_A^{(k)} = \left(d_1^{(k)}, \ldots, d_{12}^{(k)}\right) \tag{2}$$

with analogous definitions for subsets B, C, D, E. Let now denote the vectors corresponding to the kth drawing as independently obtained by the MT and the CT, by, respectively:

$$\left[\boldsymbol{d}^{(k)}\right]_{\mathrm{MT}} \text{ and } \left[\boldsymbol{d}^{(k)}\right]_{\mathrm{CT}} \tag{3}$$

with analogous subscripts (A, B, C, D, F) for the partial vectors corresponding to the various subspaces. In other words, (3) denote the kth datapoint as determined by the MT or the CT.

We can consider the "distance" between the two determinations (MT or CT) of a given datapoint,[3] referring to each subset of the features space. For example, for the A subset:

$$\Delta_A^{(k)} = \frac{1}{D_A} \sum_{i=1}^{D_A} \left| \left[d_i^{(k)}\right]_{\mathrm{MT}} - \left[d_i^{(k)}\right]_{\mathrm{CT}} \right| \tag{4}$$

[3] This is just an extension of the well-known "Manhattan metrics".

Table 2 Coherence coefficients of the cross-analysis, for the five different subsets of feature

Subset	γ
A	0.98
B	0.91
C	0.82
D	0.58
E	0.75

with analogous definitions for the distances in the remaining subsets, with the appropriate subscripts (*B*, *C*, *D*, *E*).

Finally, let us consider, for each subset, a "coherence coefficient" defined as (for the *A* subset, with analogous definitions for other subsets):

$$\gamma_A = \frac{1}{N_0} \sum_{k=1}^{N_0} \left(1 - \Delta_A^{(k)}\right) \tag{5}$$

where $N_0 = 750$ is the number of datapoints determined independently by the two teams, MT and CT.

Coefficient γ_A ranges from about 0.5 (the two set of datapoint, MT and CT, are randomly generated) to 1 (the two sets of data points coincide component by component), so it is a good estimator for the coherence of the analyses cross-performed by the two teams (Main Team and Crosscheck Team). Table 2 shows the coherence coefficient determined for the five subsets in Table 1.

The coherence coefficients given in Table 2 suggest that the analysis performed by the Analysts is very reliable as regards the features belonging to subsets *A* and *B* and, to a smaller extent, but still satisfactory, for subsets *C* and *E*. On the contrary, features belonging to the *D* subset give rise to a great disagreement between the MT and CT teams of Analysts, and so these features are not appropriate for the analysis.

5 Data Analysis

In this paper we propose the analysis of 1101 drawings collected from north to south of Italy, among which 197 from schools in northern Italy, 156 from central part and 748 from southern Italy.

The preliminary analysis proposed in this paper has been performed starting from the indicators introduced by Chambers (1983), improved by Finson et al. (1995) and by Ferguson and Lezotte (2020). We have also added the classification of the scientists' jobs.

A first interesting outcome arising from our analysis of data is represented by Fig. 6, which illustrated the correlation between the gender of the primary school students (i.e. Male or Female Author of the drawing, respectively indicated with MA and FA) and the gender of the Scientists (i.e. Male or Female Scientist, respectively indicated with MS and FS). The Figure shows that most of pupils represent scientists of their own gender. However, the figure shows an interesting difference between the drawing reproduced by male and female pupils. In fact, considering the whole sample (the first two columns, named Italy), almost all male students (94%) drew a male scientist, while only 6% of them drew a female scientist. A different situation can be observed in the case of drawing proposed by the female part of the whole sample of students. In fact, only 70% of them drew a female scientist while the 30% of them proposed a male scientist. These differences seem to confirm the gender question, as demonstrated in the previous research discussed at the beginning of the paper (Ferguson and Lezotte 2020; Losh et al. 2008). In fact, while almost all male pupils represent scientists as male persons, an important number of female pupils does not represent scientists of their same gender (30% of them represent male person doing science).

The comparison between the drawings belonging to different areas of Italy shows some interesting differences: as regards the drawings proposed by male authors, it is possible to observe 93% of MS and 7% of FS in north of Italy, 96% of MS and 4% of FS in centre of Italy and 94% of MS and 6% of FS in south of Italy; as regards female authors, it is possible to observe 75% of FS and 25% of MS in north of Italy, 80% of FS and 20% of MS in centre of Italy and 67% of FS and 33% of MS in south of Italy. This detailed analysis shows that the differences between representations proposed by male and female pupils are slightly greater in south of Italy compared to the other parts involved in our research project.

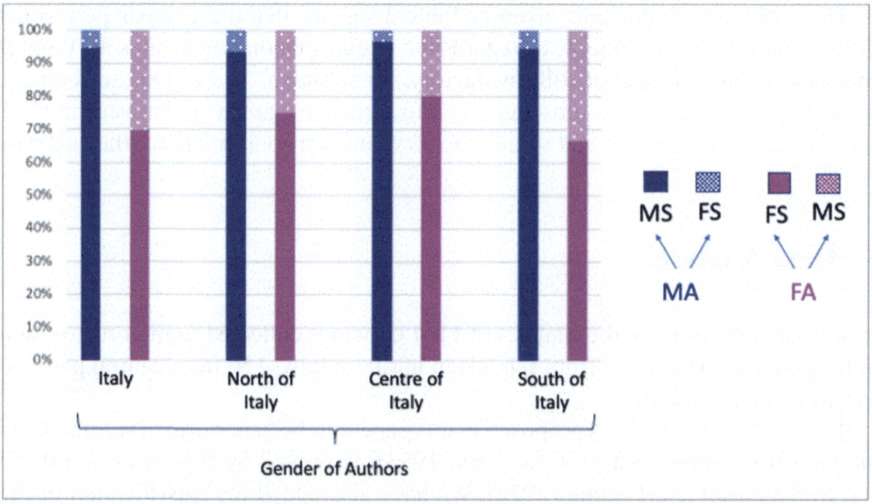

Fig. 6 The figure illustrates the relationship between the gender of the author of the drawing (i.e. Male or Female Author, respectively indicated with MA and FA) and the gender of the Scientist (i.e. Male or Female Scientist, respectively indicated with MS and FS)

A second interesting result arising from the analysis of data regards students' perception about the workplace of the scientist (Fig. 7). This classification was performed observing both the drawing of pupils and the "Information Form", where they were explicitly asked to specify the workplace of their drawn scientist in terms of indoor or outdoor job (see Fig. 1).

More than 3/4 of the enrolled pupils (77% of them) represented scientists performing indoor jobs (almost all of them working in a typical laboratory). Only 18% of pupils drew scientists working outdoor, while 5% did not specify the kind of job performed by the drawn scientist. Figure 7 shows the percentages of pupils' answers and some examples of the representation of the workplace, in terms of indoor job (orange boxes) and outdoor jobs (green boxes).

These first results (represented by Figs. 6 and 7) confirm some stereotyped representations of scientists discussed in literature, which illustrates the scientist as a male person who works in an indoor laboratory (Chambers 1983; Mead and Métraux 1957).

Another interesting result arising from our analysis regards the age of the drawn scientist (Fig. 8). This classification was carried out by taking into account some particular details. For example, a young scientist was distinguished by their hairstyle

Fig. 7 The figure illustrates (*i*) the distribution of pupils' answers regarding the workplace (i.e. indoor or outdoor) of represented scientists and (*ii*) some examples of pupils' drawings

(braids for women and short hair for men, see green boxes in Fig. 8); a neat hairstyle or the hair colour were used to characterize a middle-aged scientist (such as a chignon for women, or typical brown or blonde hair, see the blue box in Fig. 8); finally, some details were used to characterize a typical elderly scientist (such as balding man, sometimes with grey hair, see the grey box in Fig. 8).

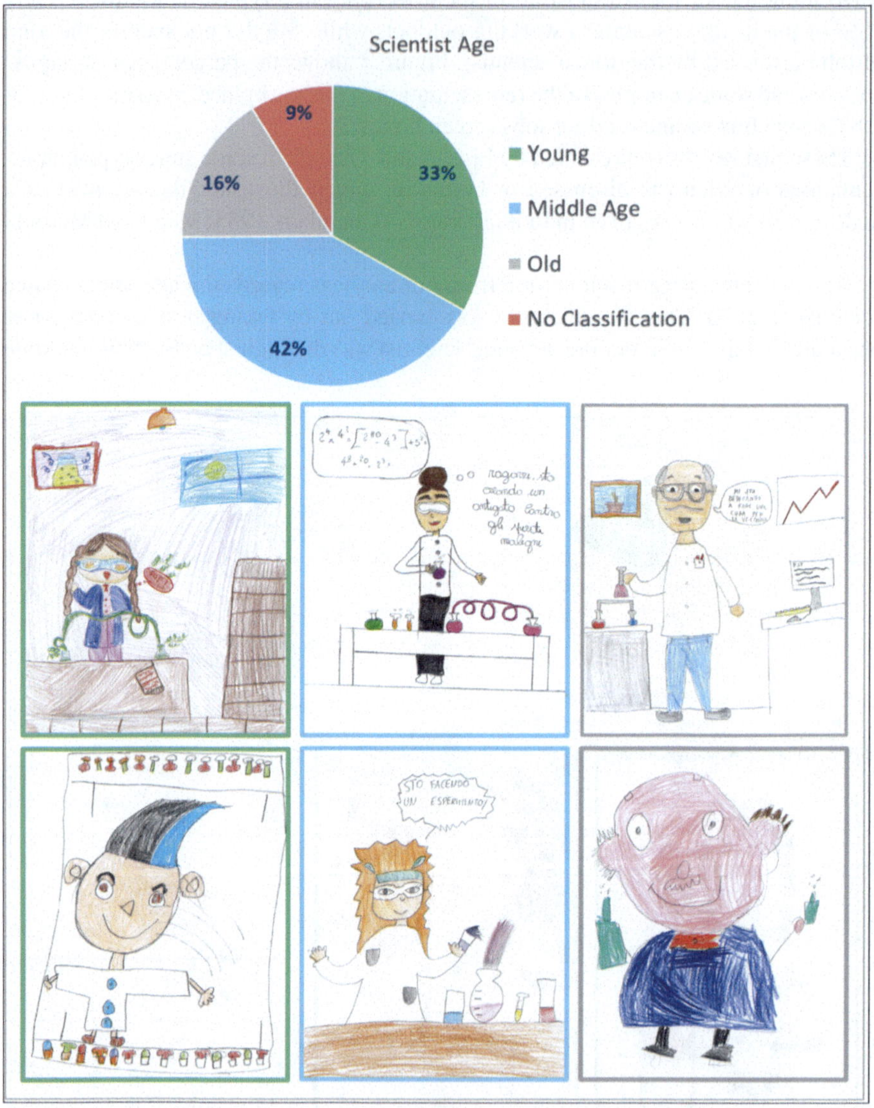

Fig. 8 The figure illustrates (i) the distribution of pupils' answers regarding the age of scientists illustrated by pupils and (ii) some examples of pupils' drawings

Figure 8 shows that almost half of students (42%) propose middle aged scientists with a neat hairstyle. This result does not confirm the stereotyped image of scientists discussed in literature of an old balding man, drawn only by 16% of enrolled pupils (Chambers 1983; Mead and Métraux 1957).

Finally, another interesting outcome regards the scientists' job (Fig. 9). Pupils drew mainly chemists (37%), biologists (14%), or medical doctors (12%). This analysis was certainly influenced by the pandemic, which has influenced our lives in the last years (starting from the beginning of 2020). In fact, in many drawings pupils explicitly wrote in the "Information Form" that the scientist drawn by them "is working on a vaccine" or "is working to solve Covid". The classification of scientists' job has been performed mainly considering the products of scientists' work (as for example, "a medicine", "the plants", "animals' life", "a vaccine", "experiments on heat and temperature").

An interesting consideration concerns the drawings of scientists who work in the field of mathematics (Fig. 9). In fact, this classification has been performed through the explicit presence of mathematical formulas (such as an algebraic expression). However, even in these cases, a lab bench with test tubes appears. This seems to indicate that all scientists often have to deal with a lab bench with test tubes, regardless of their field of work.

6 Conclusions

In this paper, we have described a nationwide study carried out on the perception, in primary school children, of science and those who practice it. The main goal of this study has been to identify and characterize the features (indicators) used by children to represent scientists and their work. The wide sample of enrolled primary school students has been chosen with the aim of analysing the different results arising from various national geographical areas (north, centre, south) and of comparing them with the results of similar surveys conducted in other parts of the world.

In this paper, we have presented a preliminary analysis of about 60% of the collected drawings performed by pupils. Our preliminary analysis has been performed starting from the indicators introduced by Chambers (Chambers 1983), improved by Finson et al. (1995) and by Ferguson and Lezotte (2020). Also in our study, the enrolled pupils propose some stereotyped representations of scientists discussed in literature, illustrating the scientist mainly as a male person who works in an indoor laboratory (Chambers 1983; Mead and Métraux 1957).

The preliminary analysis does not confirm the stereotype of an old balding man. In fact, about half of pupils represent middle-aged scientists with a neat hairstyle (like a chignon).

The future improvements of this nationwide study take into account several research goals. The first one consists in the analysis of the whole sample, considering also other indicators proposed in literature. The next step is to develop a procedure of

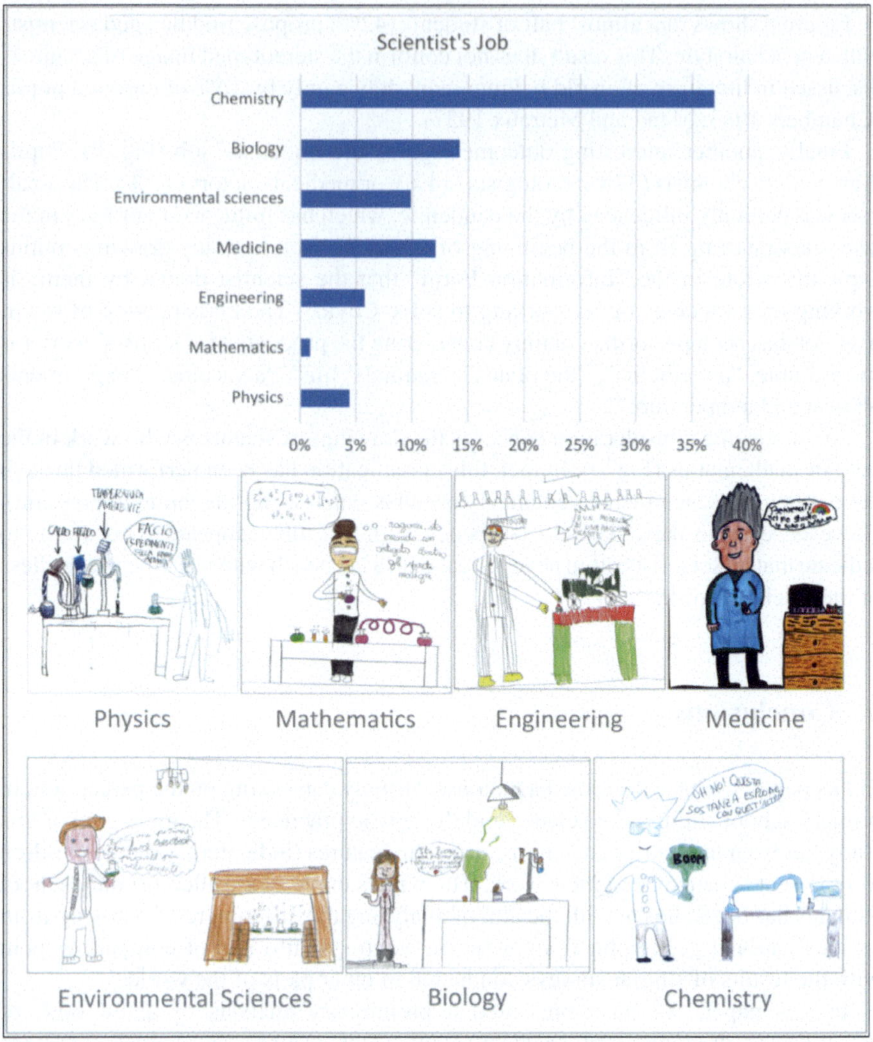

Fig. 9 The figure illustrates (i) the distribution of pupils' answers regarding the job of scientists illustrated by pupils and (ii) some examples of pupils' drawings. The classification of mathematics job has been performed through the explicit presence of mathematical formulas (such as an algebraic expression). Also in these cases, a lab bench with test tubes appears. This seems to indicate that all scientists often have to deal with a lab bench with test tubes, regardless of their field of work

analysis based on Machine Learning techniques (as for example, Cl.A., PCA, etc.), with the aim of identifying the most significant indicators, correlations, etc.

It is known from literature that empirical research based on children's drawings could be affected by limitations that still need to be addressed and clarified (Fabris et al. 2023). These limitations include the psychometric properties of tools used for the quantitative analysis of drawings, the influence of developmental levels in graphic ability, as well as cultural factors. In particular, we have observed in our sample important qualitative differences between drawings done by children belonging to early years or to late years of primary school. In this perspective, a deeper analysis on the possible influence of these factors is on the way and will be presented in a more specific paper.

Finally, it is worth observing that the findings discussed in this paper could inspire educational policies aimed to address the stereotypes confirmed by our study, in particular the gender stereotype that frames practicing science as predominantly male. These possible policies, in our opinion, could include: (i) the introduction of diverse role models showcasing women in science; (ii) the promotion of hands-on, inclusive science activities, encouraging collaboration and participation from all students, regardless of gender; (iii) the education of teachers against unconscious biases and stereotypes in their teaching practices; (iv) the diffusion of gender-balanced language in teaching materials, including visual representations. These interventions, implemented consistently over time, could help create a more inclusive environment that encourages all children to see themselves as potential scientists, regardless of gender.

References

Bezdec JC (2021) Elementary cluster analysis: four basic methods that (usually) work. River Publishers, Gistrup, DK. (In particular Ch. 5 "Feature analysis")

Bozzato P, Fabris MF, Longobardi C (2021) Gender, stereotypes and grade level in the draw-a-scientist test in Italian schoolchildren. Int J Sci Educ 43(16):2640–2662. https://doi.org/10.1080/09500693.2021.1982062

Chambers DW (1983) Stereotypic images of the scientist: the draw-a-scientist test. Sci Educ 67(2):255–265. https://doi.org/10.1002/sce.3730670213

Fabris MA, Lange-Küttner C, Shiakou M and Longobardi C (2023) Editorial: children's drawings: evidence-based research and practice. Front Psychol 14:1250556. https://doi.org/10.3389/fpsyg.2023.1250556

Farland-Smith D (2009) How Does culture shape students' perceptions of scientists? Cross-American and Chinese elementary students. J Elem Sci Educ 21(4):23–42

Farland-Smith D (2012) Development and field test of the modified draw-a-scientist test and the draw-a-scientist rubric. Sch Sci Math 112(2):109–116. https://doi.org/10.1111/j.1949-8594.2011.00124.x

Ferguson SL, Lezotte SM (2020) Exploring the state of science stereotypes: systematic review and meta-analysis of the draw-a-scientist checklist. Sch Sci Math 120(1):55–65. https://doi.org/10.1111/ssm.12382

Finson KD (2003) Applicability of the DAST-C to the images of scientists drawn by students of different racial groups. J Elem Sci Educ 15(1):15–26

Finson KD (2012) Drawing a scientist: what we do and do not know after fifty years of drawings. Sch Sci Math 102(7):335–345

Finson KD, Beaver JB, Cramond BL (1995) Development and field test of a checklist for the draw-a-scientist test. Sch Sci Math 95(4):195–205. https://doi.org/10.1111/j.1949-8594.1995.tb15762.x

Hill C, Corbett C, St. Rose A (2010) Why so few? women in science, technology, engineering, and mathematics. AAUW, American Association of University Women

Lane KA, Goh JX, Driver-Linn E (2012) Implicit science stereotypes mediate the relationship between gender and academic participation. Sex Roles 66(3–4):220–234. https://doi.org/10.1007/s11199-011-0036-z

Losh SC, Wilke R, Pop M (2008) Some methodological issues with "Draw a Scientist Tests" among young children. Int J Sci Educ 30(6):773–792. https://doi.org/10.1080/09500690701250452

Mead M, Métraux R (1957) Image of the scientist among high-school students. A Pilot Study. Science 126(3270):384–390. https://doi.org/10.1126/science.126.3270.384

Miller JD (2012) The sources and impact of civic scientific literacy. In: Bauer MW, Shukla R, Allum N (eds) The culture of science: how the public relates to science across the globe. Routledge, pp 217–240

Miller JD (2016) Civic Scientific Literacy in the United States in 2016

Miller PH, Blessing JS, Schwartz S (2006) Gender differences in high-school students' views about science. Int J Sci Educ 28(4):363–381. https://doi.org/10.1080/09500690500277664

Miller DI, Eagly AH, Linn MC (2015) Women's representation in science predicts national gender-science stereotypes: evidence from 66 nations. J Educ Psychol 107(3):631–644. https://doi.org/10.1037/edu0000005

Miller DI, Nolla KM, Eagly AH, Uttal DH (2018) The development of children's gender-science stereotypes: a meta-analysis of 5 decades of U.S. Draw-a-scientist studies. Child Dev 89(6):1943–1955. https://doi.org/10.1111/cdev.13039

Reinisch B, Krell M, Hergert S, Gogolin S, Krüger D (2017) Methodical challenges concerning the draw-a-scientist test: a critical view about the assessment and evaluation of learners' conceptions of scientists. Int J Sci Educ 39(14):1952–1975. https://doi.org/10.1080/09500693.2017.1362712

Sapia P (2019) Educazione civica e alfabetizzazione scientifica. L'ECO Della Scuola Nuova 252–253:12–13. https://fnismcs.files.wordpress.com/2020/05/educazione-civica-e-alfabetizzazione-scientifica-peppino-sapia-2-1.pdf

Shen BSP (1975) Science literacy-public understanding of science is becoming vitally needed in developing and industrialized countries alike. Am Sci 63(3):265–268

Symington D, Spurling H (1990) The 'Draw a Scientist Test': interpreting the data. Res Sci Technol Educ 8(1):75–77. https://doi.org/10.1080/0263514900080107

Turiman P, Omar J, Daud AM, Osman K (2012) Fostering the 21st century skills through scientific literacy and science process skills. Procedia-Soc Behav Sci 59:110–116. https://doi.org/10.1016/j.sbspro.2012.09.253

A Didactic Pathway on the Concept of Energy in Primary School: Cognitive Well-Being and Self-efficacy Based on Gender

Giusy Giarratano, Onofrio Rosario Battaglia, Giulia Termini, and Claudio Fazio

Abstract This research aims to investigate the gender gap in physics learning processes across five primary school classrooms. We examine the literature about the gender gap in learning processes and self-efficacy and design a questionnaire on self-efficacy and approaches to scientific disciplines, designed for primary school, emphasizing the inequalities seen between male and female students. Moreover, a didactic pathway regarding the concept of energy, utilizing the IBSE methodology, is proposed and implemented with ten-year-old students. The questionnaire was administered prior to and following the didactic pathway to assess variations in students' self-efficacy and to analyze differences between female and male participants.

Keywords Gender differences · Physics education · Primary school · Self-efficacy

1 Introduction

Despite the many strides women have made in recent decades, they remain underrepresented in some STEM fields (Corbett and Hill 2015; Gruber et al. 2021; Hill et al. 2010). Even today we can observe the gender gap in different areas of daily life. It is a problematic phenomenon that affects various areas, including school and work, with the negative result of affecting the wealth, both economic and cultural, of the nation themselves.

Women remain underrepresented in science, technology, engineering, and mathematics: they are only 35% of students in STEM education, occupy less than a third of positions in the tech sector, and only 22% of jobs in the field of artificial intelligence (UNESCO 2023).

O. R. Battaglia · G. Termini · C. Fazio
Department of Physics and Chemistry - Emilio Segrè, University of Palermo, Palermo, Italy

G. Giarratano (✉)
Department of Psychological, Physical Exercise and Training Sciences, University of Palermo, PedagogicalPalermo, Italy
e-mail: Giusy.giarratano@unipa.it

I. Testa et al. (eds.), *Connecting Physics Education Research and Practice*,
Challenges in Physics Education, https://doi.org/10.1007/978-3-031-86609-8_2

The National Science Foundation (Hamrick 2021) documents an increasing number of women graduates in STEM fields over the past two decades, particularly in the engineering sector in the United States. However, in 2020, women were underrepresented among graduates at all levels in Physical and Earth Sciences, Mathematics and Computer Science, and Engineering. In contrast, they earned 66% of bachelor's degrees, 67% of master's degrees, and 60% of doctorates in social and behavioral sciences. Among the fields in which women are most underrepresented are physics and computer science, with 24% and 21% of women graduates, respectively, in 2020 (National Center for Science and Engineering Statistics (NCSES) 2023).

In Europe, the She Figures 2021 report shows that women continue to be overrepresented in education (67%) but choose less frequently to enroll in STEM courses (Commissione europea, Direzione generale della Ricerca e dell'innovazione 2021). Data from 2018 show that women continue to be underrepresented even among Ph.D. students in the narrow STEM fields of Physical Sciences (38.4%), Mathematics and Statistics (32.5%), ICT (20.8%), Engineering (27%), and Architecture and Construction (37.2%). Between 2015 and 2018, there was little progress toward female representation among doctoral students in these narrow fields of STEM at the European level. In Italy, one of the most comprehensive reports on the university and work situation in our country is the AlmaLaurea Report. In 2022, the Report "Graduates: choices, experiences and professional achievements" was published (AlmaLaurea 2022). It is a thematic report, dedicated to outlining the analysis of gender differences in training choices and employment outcomes.

The Report makes it clear, once again, that gender is one of the factors influencing educational choices and determines different employment performance between women and men.

In Italy in 2019, female graduates accounted for 56.9% (MUR-USTAT 2021a), while the female component of those who complete a Ph.D. course is 51.1% (MUR-USTAT 2021b). Women therefore represent more than half of graduates in Italy, but their presence in the various university courses is not homogeneous.

In the 2018–2019 academic year, 18% of women enrolled in STEM fields. However, we can see some progress, as in the 2020/21 academic year, 21% of women enrolled in STEM degree courses, twice as many men (42%) (MUR-USTAT 2021c). This result is entirely in line with AlmaLaurea data, according to which, among female graduates in 2020, 18.9% obtained a degree in STEM courses, among men it is 39.2% (AlmaLaurea 2024). In the academic year 2018/2019, the presence of girls ranges from 82% in the Literary, Philosophical, Artistic, and Historical group (including the degree courses in Conservation of Cultural Heritage) to 20% in the Electronic and Information Engineering group (Rielaborazione Osservatorio Talents Venture su dati Anagrafe degli Studenti 2020). From 2017 to 2022, 70.5% of men graduated in a master's degree course in physics, while 29.5% had female graduates (AlmaLaurea 2024).

Therefore, although progress has been made toward gender equality in the life and social sciences, some sectors of STEM disciplines, such as the physical sciences, continue to be characterized by persistent gender disparities in representation, feelings of inclusion, and pay equity (Schmader 2022).

2 The Influence of Gender Stereotypes

The reasons for the lack of girls in science are certainly not due to a lack of academic skills but to socio-cultural factors (Hill et al. 2010): women's relatively minor interest in male-dominated STEM careers is likely limited by gender stereotypes (Schmader 2022).

These cultural stereotypes subtly shape women's ability to feel a sense of fit and belonging in STEM by influencing people's perceptions (of themselves and others), preferences (for activities and values), and career goals (Schmader 2022). Research has also shown that stereotypes can implicitly shape discriminatory behaviors and attitudes toward women in STEM disciplines (Rice and Barth 2016).

According to Cuddy et al. (2015), "Gender stereotypes are ideas, deposited in the storerooms of our memory, influenced by the dominant values of cultural contexts that impose a sexual analysis of reality and bring out inequalities between males and females".

Some children begin to learn these gender stereotypes as early as preschool (Gonzalez et al. 2021), within the family and social context of reference: just think of parents' expectations and the predominant role of social media. Subsequently, they can find reinforcement within the education system through books, curricular activities and teachers' behavior (Mustapha 2014).

Students' beliefs about their skills, values, and behaviors are strongly influenced by the significant figures they meet throughout their lives, and teachers are certainly among them; they can confirm or disconfirm a child's behavior based on gender expectations (Trautner et al. 2005). How a teacher interacts with a student is based on his or her personal beliefs about how a boy or girl should grow up.

Research has shown that teachers' expectations of both male and female students are influenced by gender stereotypes as early as primary school (Keller 2001; Tiedemann 2002) and that teachers tend to exhibit gender biases that might discourage girls from choosing STEM subjects (Hand et al. 2017).

Some international studies have indicated that teachers overestimate girls' language skills (Hinnant et al. 2009; Ready and Wright 2011), while boys are considered to be better at math (Tiedemann 2002), although they achieve the same levels of success. The analysis conducted by Hand, Rice, and Greenle revealed significant gender role biases: on average, teachers reported that boys are better at STEM subjects and that girls are better at humanities (Hand et al. 2017).

Scientific literature confirms that this has repercussions on students: teachers' expectations influence preferences about school subjects and pupils' future aspirations (Van den Broeck et al. 2020). Therefore, children may end up expressing professional interests in line with previously learned gender stereotypes, drastically orienting their future career choices (Ramaci et al. 2017; Zysberg and Berry 2005).

3 Self-efficacy

The pressure, conscious or not, of these prejudices could lead female students to experience the study of scientific disciplines in a negative way, with little interest, lower self-esteem, and self-efficacy.

Self-efficacy is defined as a person's confidence in their ability to acquire new information or to complete a task or activity at a prescribed level of performance (Bandura 1977). According to Bandura (1977), this concept expresses an individual's self-confidence in completing a specific task successfully.

Kurbanoğlu (2004) suggests that self-efficacy, as one of the most important variables affecting human behaviors, reflects a person's beliefs in their abilities to accomplish a mission. The confidence that students have in their own abilities can help determine what to do with the knowledge and skills they possess (Tatar et al. 2009).

Bandura (1977) theorized that self-efficacy beliefs are formed through the processing of four sources: Mastery Experience, Vicarious Experience, Verbal Persuasion, and Emotional State.

Educational studies have shown that self-efficacy beliefs are highly correlated with academic performance through their influence on persistence, effort, motivation, and choice (Archer et al. 2013; PISA 2019).

A wealth of literature exists that highlights the importance of self-efficacy (SE) for learning and subsequent academic performance (Honicke and Broadbent 2016). According to Bandura's cognitive theory (Bandura 1977), self-efficacy motivates students to engage in academic tasks and encourages them to maintain motivation in the face of failures or temporary setbacks. Zimmerman (2008) demonstrated that self-efficacy is one of the main predictors of intrinsic motivation, which is motivation driven by the pleasure of learning for its own sake rather than for external rewards. Research explores how self-efficacy influences academic outcomes and is closely linked to motivation and students' overall well-being, highlighting a clear connection between these factors (Honicke and Broadbent 2016).

Students usually make decisions according to their self-efficacy beliefs and choose educational pathways and careers in areas they deem themselves capable of succeeding. Students with low self-efficacy beliefs for science would therefore be less likely to choose careers within science. Indeed, girls are less likely to aspire to science careers than boys (Webb-Williams 2018) and PISA (2019) results show gender gaps in drive, motivation and self-belief, with girls reporting lower self-belief in their abilities. Girls tend to underestimate their own performance (Webb-Williams 2018).

Science Self-Efficacy (SSE) can be described as an individual's self-confidence in completing tasks successfully in science (Carrol et al. 2023). Students who show a low belief in self-efficacy for science would be less likely to choose careers within science (Webb-Williams 2018). This highlights the importance of developing a good sense of self-efficacy.

Gender differences in self-efficacy sources have been studied and the results tend to vary depending on the study domain. In the field of mathematics, Joët et al. (2011) reported that boys in France have higher mastery, received more positive social messages, and felt less anxious than girls (Butz and Usher 2015). The opposite trend was found for writing, with girls having higher levels of mastery and lower levels of anxiety (Pajares et al. 2007). Furthermore, during science lessons, girls tend to participate less than boys (Gorard and See 2009). Despite the data reported in these studies, the picture emerging in the literature is not consistent. In the scientific field, Britner and Pajares (2006) found that American middle school girls (from fifth to eighth grade) had stronger self-efficacy beliefs than boys. Additionally, some studies have found that girls use higher levels of vicarious experience and social persuasion than boys and employ more self-regulation strategies in mathematics (Cleary and Chen 2009).

Therefore, the research provides evidence that gender differences exist in the formation of self-efficacy, but the patterns are inconsistent. The international landscape in literature presents a very complex picture.

In many studies, gender differences in attitudes toward science are very evident, particularly from secondary school through to university (AlmaLaurea 2024). Specifically, studies highlight the difficulties women face in developing a scientific identity and a sense of belonging in the scientific community (Schmader 2022). On the other hand, this phenomenon is not true in all contexts and for all women. The study conducted by Reid and Skryabina (2003) explored the attitudes and perceptions of Scottish girls and boys toward physics, in the age range of 10–18 years, highlighting how these attitudes are very positive, especially among girls.

In summary, gender differences in science are evident but in different modes and forms. There is a rich body of literature on this topic that seeks to shed light on the various aspects and variables involved, particularly the relationship with self-efficacy and well-being.

According to the OECD, self-efficacy is one of the variables of the cognitive dimension closely related to the concept of well-being (OECD 2017). When students perceive themselves as competent in learning, they experience less stress and anxiety, factors that often undermine school well-being (Caprara et al. 2008). Therefore, factors such as self-efficacy, motivation, and well-being are closely interconnected in learning and profoundly influence students' academic success and development. These three aspects reinforce each other, creating a virtuous cycle that can promote more effective and fulfilling learning. The OECD defines well-being as a dynamic state characterized by students experiencing the ability and opportunity to fulfill their personal and social goals (Borgonovi and Pàl 2016).

Students' well-being is an objective that the school must set as a priority, since it is increasingly considered as an indicator of the quality of the teaching–learning process.

The well-being concept is complex, as it englobes a variety of elements like physical and mental health, happiness, and satisfaction of student's lives (Govorova et al. 2020).

Among the PISA questionnaires that are administered to students, one in particular is focused on well-being: it covered a wide range of variables aimed at collecting information about students' life satisfaction, social connections and activities, subject well-being, and family support. In addition to the PISA questionnaire, there are different physics self-efficacy's surveys for high school or University but there is an absence of a suitable instrument to measure the science self-efficacy of children.

During pre-adolescence (10–14 years old) is a key time when children form their attitudes and beliefs about science (Archer et al. 2013) and when their self-efficacy beliefs are at their most malleable (Lazowski and Hulleman 2016). This is also often the time at which students' interest in science begins to decline (Sheldrake and Mujtaba 2020). Despite this, most SSE research has focused on students aged 13 years or older (Chen and Usher 2013; Lin and Tsai 2018), with relatively fewer studies focusing on younger students (Dorfman and Fortus 2019; Webb-Williams 2018).

Studies (Hendrickson 2019; Pajares and Schunk 2002) suggest that, when students are actively involved in the construction of knowledge, as is the case with IBE, they develop a sense of competence and control over their learning abilities, thus increasing their self-efficacy. The study conducted by Fauth et al. (2019) investigates how active methodologies, such as inquiry-based learning, influence students' perceptions of teaching quality and the impact on their self-efficacy and motivation. The study found that this approach promotes self-efficacy, as students feel more competent in solving complex problems through the autonomous application of critical and analytical thinking strategies.

4 The Research

4.1 Research Purpose and Questions

This research aims to study primary pupils' promotion of a sense of well-being and self-efficacy on scientific disciplines, with particular attention to gender differences, through an Inquiry-Based-Education didactic pathway focusing on the energy concept.

Underpinning this work is the intention of giving a better insight to practitioners on how students' self-efficacy may be strengthened during science lessons.

A pilot study was conducted with a small sample of students to verify the hypothesis that the Inquiry-Based Education methodology can promote the transition from common knowledge to scientific knowledge, while simultaneously encouraging the participation of all students, improving levels of enjoyment, well-being, and self-efficacy, with particular attention to gender differences.

The research questions we will answer in our whole research are the following:

1. How do pupils approach the study of science and physical content, and in particular what is their levels of self-efficacy and well-being in learning?
2. To what extent can an educational intervention based on Inquiry Based Education promote an improvement in well-being, self-efficacy, and interest of students in the study of scientific disciplines and particularly of physics?
3. How does the gender influence the previous aspects?

4.2 Study Participants

The study sample comprises 75 5th grade students (40 boys, 35 girls) from one Sicilian primary school. A subsample of 10 students participated in semi-structured interviews (5 girls and 5 boys).

The institute "IV Circolo Girgenti" is composed of two school buildings. The main building is located in the city of Bagheria (PA), the secondary one is located in a hamlet near the city (Aspra). The users that the school welcomes are mixed; for the most part, students come from a low to medium socio-cultural background. The pupils who took part in the survey are between 10 and 11 years old and they come from a disadvantaged social and economic context.

4.3 Methodology

Promoting learning is a very complex concept: our research group has, to this end, created a summary and schematic map that highlights three main aspects of learning that can be considered relevant in the light of the research literature.

Among the possible dimensions of learning useful for characterizing the promotion of learning found in the literature (Fazio et al. 2023; Marzano 1992), with specific reference to scientific learning, we have concentrated in this research on: acquisition of conceptual knowledge, intellectual growth, development of habits of mind suitable for learning science. Moreover, we identified 13 sub-dimensions of learning that better specify the abovementioned learning dimensions (Fazio et al. 2023). Among these, in this research we focus on:

- Transition from common-sense knowledge to scientific knowledge.
- Development of reasoning skills aimed at interpreting reality and experiences.
- Well-being in learning.
- Growth mindset (Dweck 2015).
- Perception of self-efficacy

The methodology used during the pedagogical activities is based on inquiry-based science education (IBSE) (Bybee et al. 2006). IBSE is based on constructivist theories, which assert that individuals actively construct their own understanding and knowledge of the world through experiences and reflection on those experiences (Bybee et al. 2006).

IBSE refers to an active exploration process. Through this process, critical, logical, and creative skills are employed to ask questions about specific situations of interest and to engage in finding answers to those questions (Bybee 1993). It explicitly refers to the typical methods followed by scientists to conduct research, defined as an "Investigation Cycle." According to Bybee, a model consisting of 5 E's can be identified, each corresponding to a specific phase: Engage, Explore, Explain, Extend, and Evaluate.

The activities carried out during the teaching intervention were designed based on these phases. Students worked in circle time, in cooperative learning, and in pairs. They observed phenomena, proposed hypotheses that they later tested, and worked with practical and playful materials.

Just as Scientific Inquiry aims to provide answers to the questions that scientists pose, learning Science based on Scientific Inquiry aims to enable learners to construct answers to their questions through a methodology that is equally clear and rigorous (Fazio et al. 2021). The methodology used allows for the real engagement of all students, improving the effectiveness of learning and the level of enjoyment.

Through this teaching approach, we expect to encourage all students to actively participate in the construction of their own learning, motivating and engaging their curiosity. We also expect this to improve their self-efficacy in scientific disciplines.

During the activities, students pose questions, plan surveys aimed at answering them, and review what is already known in the light of experimental evidence-reflecting what scientists do.

To analyze the transition from common-sense knowledge to scientific knowledge and the development of reasoning skills aimed at interpreting reality and experiences, a questionnaire on physical contents is proposed at the beginning and at the end of the didactic pathway.

To analyze attitudes, approach and the sense of self-efficacy toward scientific disciplines, a questionnaire is administered at the beginning and at the end of the course. In addition, interviews are conducted with a small sample and guided discussions in circle time are recorded.

4.4 The Self-efficacy's Questionnaire

Our research group has designed a self-efficacy's survey for students of primary school on the basis of other studies conducted at different schooling levels. In particular, the items of the PISA questionnaire (PISA 2019) and of the IS-SEC-Q (Pajares et al. 2007) were analyzed and readjusted. The first one was carried out for Italian high school students and the second was carried out for Irish primary school students.

The source questionnaires are validated, but a new validation process was carried out in a primary school in the same geographic area of the pilot group, with 10-year-old students.

The items from the PISA questionnaire were adapted to account for the different ages of our sample compared to the age of the students (15 years old) to whom the PISA questionnaire is addressed. The items from the IS-SEC-Q questionnaire were translated into Italian.

The questionnaire preparation was followed by preliminary validation. First, an "a priori" analysis of the potential student responses to the questionnaire items was then conducted. Following Brousseau (1997), this type of analysis helps identify the expected answering strategies students might use when confronted with a problem, as well as any alternative responses that may arise. This process is highly valuable for researchers, as it helps go in depth into the questionnaire content (Foxcroft et al. 2013). In fact, anticipating possible student strategies can reveal weaknesses in the questions, allowing researchers to revise them before administering the questionnaire. This analysis is performed independently of direct observation (hence the term "a priori"), providing a reference point for the subsequent examination of students' actual responses to the questionnaire items.

To further enhance the validity of the questionnaire, the a priori analysis was carried out independently by two researchers. A consensus was then reached through discussion, resulting in a final, shared version optimized for the research objectives.

To further test the validity of the questionnaire and highlight any other issues with the questions, such as unclear or ambiguous terminology, a face-validation (Jensen 2003) involving a fifth-grade class from the same school has been conducted. Each student in the pilot group completed the questionnaire, and a focus group was then conducted with the students to clarify the meaning of their responses and obtain the final version of the tool to be used with the research sample.

The questionnaire is divided into five sections and each section contains several items. In the first four sections, the participants were asked to respond on a 5-point Likert-like scale. In the last one, they responded on an 11-point Likert-like scale (from 0 to 10).

Bandura (Kurbanoğlu 2004) recommended that to be most effective, self-efficacy instruments should be tailored to the relevant task and within a specific domain and context.

This questionnaire was specifically designed to assess the strength of students' SSE beliefs and their perceived exposure to the four sources of SSE. For example, in the second section, pupils should indicate how much they agree with phrases such as: "Science is easy for me", "I do well even the most difficult science homework", and "Even when I try hard, I do poorly in science".

Based on the literature (Carrol et al. 2023), we identified 11 variables in our questionnaire, strictly related to the sub-dimensions of learning we want to analyze in this research, that can be studied by means of the questionnaire items. To calculate the variable scores, the response scores for constituting items for a variable were summed, and then divided by the total number of items for each participant. In summary, the outcome variables presented in this work are: Desire to learn and extend, Growth mindset, General academic SE, Knowledge-specific SSE: life Sciences, Knowledge-specific SSE: physics/chemistry Sciences, Sources of SSE (examined as four latent variables: Mastery Experience, Vicarious Experience, Verbal Persuasion, Emotional State), Skills-related school SSE.

4.5 In Class Activities

The didactic pathway involved five classes of Sicilian primary school with 10-year-old students for 14 h for each class. The pathway contains activities dealing with some relevant phenomena related to energy conservation. Most of the didactic tools consisted of toys, and many of the materials were inexpensive and easy to find, as well as being familiar and engaging to students.

The structuring of activities began with a literature study on students' misconceptions and common knowledge about energy. References are made to some already validated pathways in the literature (Colonnese et al. 2012; Millar 2005; Neumann et al. 2013).

Based on the literature, it was chosen not to discuss the work concept (Colonnese et al. 2012).

Energy is associated with various everyday connotations and gets extensive media attention due to socio-economically or environmentally relevant topics such as climate change, (renewable) energy sources, health issues (ozone hole, malnutrition, nuclear energy), or globalization. As a result, students' understanding of energy is influenced by both everyday and scientific contexts (Opitz et al. 2015).

The concept of energy is addressed starting from the discussion on renewable and non-renewable energy sources, through the use of images and videos with a practical and laboratory approach, exploiting students' tendency to associate energy with living beings and movement. The activities are carried out in pairs, in cooperative learning, and in circle time. Every day, students transcribe the activities carried out in their own work notebooks, thereby verbalizing the work done both orally and in written form.

Nordine et al. (2018) states that students often struggle to see the usefulness of energy analysis for interpreting the world around them. One reason for this may be an over-reliance on the idea of energy forms in introductory energy instruction, which may unintentionally suppress, rather than prompt, insights into how and why phenomena occur. For this reason, we welcome the suggestion of Neumann et al. (2013) and agree on an instructional approach on energy that emphasizes energy transfers between systems.

The well-known tendency for pupils to conceptualize energy as a quasi-material substance that can flow from one object to another (Opitz et al. 2015; Taber 1989) may be natural, and may encourage the development of the ideas of transfer and conservation, but it is an idea that is not consistent with the contemporary view of energy held by physicists (Millar 2005; Taber 1989). Therefore, the energy concept is addressed from the idea of a state property of bodies that transforms during interactions (Opitz et al. 2015).

Some researchers (Brook and Wells 1988; Sefton 2004) point out that the energy concept in static situations often causes confusion among students. For this reason, we present energy in situations in which observable changes are taking place: wheels spinning more quickly, objects falling from higher to lower positions, temperatures increasing, etc. Moreover, according to Carr and Kirkwood (1988), the initial and final states of the systems must be clearly identified to focus students' attention on the energy types involved in the physical situation being analyzed.

The activity begins with a brainstorming session on the concept of energy, discussing all the situations in which this word is commonly used. Nowadays, students often hear about 'renewable energy' and 'non-renewable energy'. After this first distinction, reference is made to "human energy" (recognized as internal energy of the human body) which constitutes an intuitive approach for children. Food thus becomes the fuel that allows us to exert efforts. This starting point is the basis for introducing other "types" of energy: kinetic energy, potential energy, thermic energy, and energy associated with lighting. Starting from the concept of energy related to the movement of human beings, we can gradually introduce energy related to non-living beings. Initially, reference is made to the elements (sun, water, wind, heat), and then we move on to inanimate objects such as cars, bicycles, falling objects. Rotational kinetic energy is presented through a practical experience with a bike and a dynamo battery. Potential energy is demonstrated through the dropping of balls and stones of different weights into sand. Thermal energy is introduced through the temperature variation of a metal plate repeatedly struck. In addition, students will build a small water turbine using recycled materials to observe how water can power a light bulb.

The common thread of all the experiments is the transformation of energy, highlighting how energy can be transferred from one system to another.

5 Analysis of the Data

At this stage, the full analysis of the data we collected is ongoing.

The analysis of collected data will be carried out through different modalities: descriptive and significance analysis and cluster analysis (Battaglia et al. 2019) for the administered tests, using the research sample of 75 students. The students' work notebooks and audio recordings of classroom discussions will also be collected, and analyzed through thematic analysis (Braun and Clarke 2006).

The latter is a qualitative analysis method that will allow us to identify and analyze recurring themes emerging from the data. Moreover, it will enable us to explore students' ideas and experiences at a finer grain level. For the use of thematic analysis, we refer to the guidelines from previous studies (Fazio et al. 2023).

Through descriptive analysis, the average scores from the pre-test and post-test data of the girls and boys in the self-efficacy questionnaire will be compared. Then, we will compare the data between the two genders. A t-test analysis will also be conducted, and Cohen's d will be calculated. This analysis will allow us to observe any statistically significant changes in the variables identified in the questionnaire between the first and second administrations. Due to the preliminary nature of this study, the research does not aim to generalize the obtained data.

We observed that, during the administration of the self-efficacy questionnaire, the students had some difficulty understanding the meaning of two items (see items 29–30 in the Appendix). For this reason, both items have been removed in the next version of the questionnaire.

The analysis conducted will subsequently be used to guide a full-scale experiment, which will include the creation and validation of a new questionnaire on SSE. In the next experiment, a sample of 400 primary school students from different areas of Sicily is planned. Factor analysis, as well as cluster analysis, will be conducted using SPSS software and custom-coded software (Battaglia et al. 2019). This will allow us to expand on the results obtained from the pilot study and further test our hypotheses.

6 Conclusion

The existing literature provides a broad overview of gender-related issues.

Although in recent decades the gender gap in scientific disciplines has significantly narrowed, women continue to be underrepresented in many STEAM fields, such as physics and engineering (Corbett and Hill 2015; Gruber et al. 2021; Hill et al. 2010; UNESCO 2023).

Recent research partially attributes women's disengagement from these fields (i.e., their relatively lower STEM self-concept) to the role of prevalent automatic associations linking concepts related to STEM more to men than to women (i.e., implicit math-gender stereotypes) (AlmaLaurea 2022, 2024; Commissione europea, Direzione generale della Ricerca e dell'innovazione 2021; Hamrick 2021; Hill et al. 2010; MUR-USTAT 2021a, b, c; National Center for Science and Engineering Statistics (NCSES) 2023; Rielaborazione Osservatorio Talents Venture su dati Anagrafe degli Studenti 2020; Schmader 2022; UNESCO 2023).

In addition, several studies have focused on the concepts of self-efficacy and cognitive well-being, and the differences observed between males and females (Webb-Williams 2018). Indeed, women often exhibit lower self-efficacy in science, which in turn leads to lower motivation to study. It is important to emphasize that the phenomenon described here is not universal, as there are several exceptions reported in various studies (Britner and Pajares 2006; Cleary and Chen 2009; Reid and Skryabina 2003).

This study may shed light on the need to encourage and motivate female students in scientific disciplines to reduce the still existing gap between the two genders (OECD 2017; PISA 2019; Webb-Williams 2018). Moreover, primary school teachers play a fundamental role. The scientific literature confirms that teachers can influence preferences about school subjects and pupils' future aspirations (Van den Broeck et al. 2020). To promote greater gender equality in the teaching and learning of science and mathematics, as well as in the future career choices of girls, it is crucial to go beyond a traditional approach (Mizala et al. 2015).

Appendix: The Self-efficacy's Questionnaire

Age: Class:

Gender:

- Male
- Female

How well do you agree with the following?	Not at all	A little	Somewhat	A lot	Completely
I am a very curious					
I enjoy learning new things					
I think learning new things is boring					
When the tasks become difficult, I stop					
I like to know how things work					
I am a highly creative person					
I am able to think quickly					
I enjoy working with my classmates					
I worry about many things					
I feel nervous when I think about tests					

How well could you do the following…	Terribly	Poorly	Neither well nor poorly	Well	Perfectly
Italian					
Math					
Science					
English					

How well could you answer questions on the following science topics?	Terribly	Poorly	Neither well nor poorly	Well	Perfectly
Plants					
Animals					
States of matter					
Heat and temperature					
Human body					
Liquids					
Solutions and blends					
Energy					
Water cycle					
Forces					
The Universe and the Solar System					
Electricity					
Magnets					
Using science in your everyday lives					
How science has made the world better					

How well do you agree with the following?	Strongly disagree	Disagree	Neither agree nor disagree	Agree	Strongly agree
Science is one of my favorite subjects					
I received excellent grades in science tests					
I enjoy learning new science topics					
I can't wait to do science					
I lose interest during science lessons					
Science is easy for me					
I am interested in learning topics related to science					
Even when I study vary hard, I do poorly in science					
When science topics are explained to me, I understand the concepts very well					
When I see how another student does a science experiment, I would like to do it too					
I have always been successful with science					
I like science assignments					
I can usually answer questions or science tasks well					

(continued)

(continued)

How well do you agree with the following?	Strongly disagree	Disagree	Neither agree nor disagree	Agree	Strongly agree
I am unable to think clearly when doing science work					
I learn science quickly					
I do well even on the more difficult science assignments					
There are people who do poorly in science, no matter how much they study					
My parents tell me that I am good at science					
I feel anxious when I have to do science					
When I see my teacher doing science experiments, I imagine doing them myself					
My teacher has told me that I am good at learning science					
My classmates like to work with me in science because they think I'm good at it					

How capable do you think you are at performing these activities?	0	1	2	3	4	5	6	7	8	9	10
Connecting science topics with each other											
Explaining science topics to my classmates if they haven't understood something											

(continued)

(continued)

How capable do you think you are at performing these activities?	0	1	2	3	4	5	6	7	8	9	10
Working together with my classmates to better understand science topics											
Write a good report about an experiment											
Correctly use instruments during an experiment											
Encouraging my classmates to participate in discussions											
Conducting science experiments using practical materials											

References

AlmaLaurea (2022) Laureate e laureati: scelte, esperienze e realizzazioni professionali. Rapporto 2022. https://www.almalaurea.it/i-dati/le-nostre-indagini/indagini-tematiche/laureate-e-laureati-scelte-esperienze-e-realizzazioni-professionali

AlmaLaurea (2024). https://www2.almalaurea.it/cgiasp/classi/Scheda.aspx?coceAggr=11020&tipoCorso=Ls&LANG=it

Archer L, Osborne J, DeWitt J, Dillon J, Wong B, Willis B (2013) ASPIRES: Young people's science and career aspirations, age 10–14. King's Collage Londono, Londono

Bandura A (1977) Self-efficacy: toward a unifying theory of behavioral change. Psychol Rev 84(2):191–215

Battaglia OR, Di Paola B, Fazio C (2019) Unsupervised quantitative methods to analyze student reasoning lines: theoretical aspects and examples. Phys Rev Phys Educ Res 15(2):020112

Borgonovi F, Pàl J (2016) A framework for the analysis of student well-being in the PISA 2015 study: being 15 in 2015. In: OECD Education Working Papers. OECD Publishing Paris, France

Braun V, Clarke V (2006) Using thematic analysis in psychology. Qual Res Psychol 3(2):77–101

Britner SL, Pajares F (2006) Sources of science self-efficacy beliefs of middle school students. J Res Sci Teach 43:485–499

Brook AJ, Wells P (1988) Conserving the circus? An alternative approach to teaching and learning about energy. Phys Educ 23:80–85

Brousseau G (1997) In: Cooper M, Balacheff N, Sutherland R, Warfield V (eds) Theory of didactical situations in mathematics. Kluwer Academic, Dordrecht, The Netherlands

Butz AR, Usher EL (2015) Salient sources of self-efficacy in reading and mathematics. Contemp Educ Psychol 42:49–61

Bybee RW (1993) Reforming science education. Social perspectives & personal reflections. Teachers College Press, 1234 Amsterdam Avenue, New York, NY, 10027 (Hardcover: ISBN-0-8077-3261-3; paperback: ISBN-0-8077-3260-5)

Bybee R, Taylor J, Gardner A, Scotter P, Carlson J, Westbrook A, Landes N (2006) The BSCS 5E instructional model: origins, activeness, and applications. BSCS 1–66

Caprara GV, Fida R, Vecchione M, Del Bove G, Vecchio GM, Barbaranelli C, Bandura A (2008) Longitudinal analysis of the role of perceived self-efficacy for self-regulated learning in academic continuance and achievement. J Educ Psychol 100(3):525

Carr M, Kirkwood V (1988) Teaching and learning about energy in New Zealand secondary school junior science classrooms. Phys Educ 23:86–91

Carrol S, McCauley V, Grenon M (2023) Science self-efficacy beliefs of upper primary students in Ireland. Int J Sci Educ 1–21

Chen JA, Usher EL (2013) Profiles of the sources of science self-efficacy. Learn Individ Differ 24:11–21. https://doi.org/10.1016/j.lindif.2012.11.002
Cleary T, Chen P (2009) Self-regulation, motivation, and math achievement in middle school: variations across grade level and math context. J Sch Psychol 47:291–314
Commissione europea, Direzione generale della Ricerca e dell'innovazione (2021) She figures 2021: gender in research and innovation: statistics and indicators, Publications Office. https://data.europa.eu/doi/10.2777/06090
Colonnese D, Heron PR, Michelini M, Santi L, Stefanel A (2012) A vertical pathway for teaching and learning the concept of energy. Rev Sci Math ICT Educ 6:21–50
Corbett C, Hill C (2015) Solving the equation: the variables for women's success in engineering and computing. American Association of University Women, Washington, DC
Cuddy AJC, Wolf EB, Glick P, Crotty S, Chong J, Norton MI (2015) Men as cultural ideals: cultural values moderate gender stereotype content. J Pers Soc Psychol 109(4):622–635
Dorfman BS, Fortus D (2019) Students' self-efficacy for science in different school systems. J Res Sci Teach 56(8):1037–1059. https://doi.org/10.1002/tea.21542
Dweck C (2015) Carol Dweck revisits the growth mindset. Educ Week 35(5):20–24
Fauth B, Decristan J, Rieser S, Klieme E, Büttner G (2019) Student ratings of teaching quality in primary school: dimensions and prediction of student outcomes. Learn Instr 60:1–9; Mackay J, Parkinson J (2010) Gender, self-efficacy and achievement among South African technology teacher trainees. Gend Educ 22(1):87–103
Fazio C, Carpineti M, Faletič S, Giliberti M, Jones G, Mcloughinn E et al (2021) Strategies for active learning to improve student learning and attitudes towards physics. In: Teaching-learning contemporary physics: from research to practice. Springer International Publishing, Cham, pp 213–233
Fazio C, Gallitto AA, Galiano CG, Giarratano G, Grazia I, Termini G, Battaglia OR (2023) An approach to research-based design of teaching-learning sequences in the context of physics education: theoretical frameworks, pedagogical methods, and examples of data analysis. Il Nuovo Cimento 46:199–227
Foxcroft C, Paterson H, Le Roux N, Herbst D (2013) Psychological assessment in South Africa: a needs analysis, the test use patterns and needs of psychological assessment practitioners. HSRC Press, Pretoria, South Africa, 2004. http://intranet.hsrc.ac.za/research/output/outputDocuments/1716_Foxcroft_Psychologicalassessmentin%20SA.pdf
Gonzalez AM, Odic D, Schmader T, Block K, Baron AS (2021) The effect of gender stereotypes on young girls' intuitive number sense. PloS ONE 16(10):e0258886
Gorard S, See BH (2009) The impact of socio-economic status on participation and attainment in science. Stud Sci Educ 45(1):93–129
Govorova E, Benìtez I, Muniz J (2020) Predicting student well-being: network analysis based on PISA 2018. Int J Environ Res Publ Health 17(11):4014
Gruber J, Mendle J, Lindquist KA, Schmader T, Clark LA et al (2021) The future of women in phychological science. Perspect Psychol Sci 16(3):483–516
Hamrick K (2021) Women, minorities, and persons with disabilities in science and engineering. Rep., Natl. Sci. Found., Alexandria, VA. https://ncses.nsf.gov/pubs/nsf21321/report
Hand S, Rice L, Greenlee E (2017) Exploring teachers' and students' gender role bias and students' confidence in STEM fields. Soc Psychol Educ 20:929–945. https://doi.org/10.1007/s11218-017-9408-8
Hendrickson P (2019) Effetto delle tecniche di apprendimento attivo sull'entusiasmo, l'interesse e l'autoefficacia degli studenti. J Polit Sci Educ 17(2):311–325. https://doi.org/10.1080/15512169.2019.1629946
Hill C, Corbett C, St. Rose A (2010) Why so few? Women in science, technology, engineering, and mathematics. American Association of University Women, Washington, DC
Hinnant JB, O'Brien M, Ghazarian SR (2009) The longitudinal relations of teacher expectations to achievement in the early school years. J Educ Psychol 101(3):662

Honicke T, Broadbent J (2016) The influence of academic self-efficacy on academic performance: a systematic review. Educ Res Rev 17:63–84
Jensen M (2003) Questionnaire validation: a brief guide for readers of the research literature. Clin J Pain 19:345; Anastasi A (1988) Psychological testing. Macmillan, New York, p 144
Joët G, Usher EL, Bressoux P (2011) Sources of self-efficacy: an investigation of elementary school students in France. J Educ Psychol 103(3):649–663
Keller C (2001) Effect of teachers' stereotyping on students' stereotyping of mathematics as a male domain. J Soc Psychol 141:165–173
Kurbanoğlu SS (2004) Self efficacy belief and its importance for information professionals. Bilgi Dünyası 5(2):137–152
Lazowski RA, Hulleman CS (2016) Motivation interventions in education: a meta-analytic review. Rev Educ Res 86(2):602–640. https://doi.org/10.3102/0034654315617832
Lin TJ, Tsai CC (2018) Differentiating the sources of Taiwanese high school students' multidimensional science learning self-efficacy: an examination of gender differences. Res Sci Technol Educ 48:575–596
Marzano RJ (1992) A different kind of classroom: Teaching with dimensions of learning. Association for Supervision and Curriculum Development, 1250 North Pitt Street, Alexandria, VA 22314
Millar R (2005) Teaching about energy. Department of Educational Studies, Research Paper 2005/11. York University, York
Mizala A, Martínez F, Martínez S (2015) Preservice elementary school teachers' expectations about student performance: how their beliefs are affected by their mathematics anxiety and student's gender. Teach Teach Educ 50:70–78. https://doi.org/10.1016/j.tate.2015.04.006
MUR-USTAT (2021a) Laureati. https://dati.ustat.miur.it/dataset/laureati
MUR-USTAT (2021b) Dottori di ricerca. https://dati.ustat.miur.it/dataset/formazione-post-laurea
MUR-USTAT (2021c) Focus "Le carriere femminili in ambito accademico". https://ustat.miur.it/media/1197/focus_carrierefemminili_universit%C3%A0_2021.pdf
Mustapha A (2014) Sex roles in English language textbooks in Nigerian schools. J ELT Appl Linguist (JELTAL) 2(2):69–81
National Center for Science and Engineering Statistics (NCSES) (2023) Diversity and STEM: Women, Minorities, and Persons with Disabilities 2023. Special Report NSF 23-315. National Science Foundation, Alexandria, VA
Neumann K, Viering T, Boone WJ, Fischer HE (2013) Towards a learning progression of energy. J Res Sci Teach 50(2):162–188
Nordine J, Fortus D, Lehavi Y, Neumann K, Krajcik J (2018) Modelling energy transfers between systems to support energy knowledge in use. Stud Sci Educ 54(2):177–206. https://doi.org/10.1080/03057267.2018.1598048
OECD (2017) PISA 2015 Results (Volume III). OECD, Paris, France
Opitz ST, Harms U, Neumann K et al (2015) Students' energy concepts at the transition between primary and secondary school. Res Sci Educ 45:691–715. https://doi.org/10.1007/s11165-014-9444-8
Pajares F, Schunk DH (2002) Self and self-belief in psychology and education: a historical perspective. In: Improving academic achievement. Academic Press, pp 3–21
Pajares F, Johnson MJ, Usher EL (2007) Sources of writing self-efficacy beliefs of elementary, middle, and high school students. Res Teach Engl 42:104–120 (Jai Press, Greenwich, CT)
PISA O (2019) Results (Volume II): where all students can succeed. PISA
Ramaci T, Pellerone M, Ledda C, Presti G, Squatrito V, Rapisarda V (2017) Gender stereotypes in occupational choice: a cross-sectional study on a group of Italian adolescents. Psychol Res Behav Manag 10:109
Ready DD, Wright DL (2011) Accuracy and inaccuracy in teachers' perceptions of young children's cognitive abilities: the role of child background and classroom context. Am Educ Res J 48(2):335–360
Reid N, Skryabina EA (2003) Gender and physics. Int J Sci Educ 25(4):509–536

Rice L, Barth JM (2016) Hiring decisions: the effect of evaluator gender and gender stereotype characteristics on the evaluation of job applicants. Gend Issues 33(1):1–21
Rielaborazione Osservatorio Talents Venture su dati Anagrafe degli Studenti (2020)
Schmader T (2022) Annual review of psychology: gender inclusion and fit in STEM
Sefton IM (2004) Understanding energy. In: Proceedings of 11th biennial science teachers' workshop. University of Sydney, Sydney. http://sydney.edu.au/science/uniserve_science/school/curric/stage6/phys/stw2004/sefton1.pdf
Sheldrake R, Mujtaba T (2020) Children's aspirations towards science-related careers. Can J Sci Math Technol Educ 20(1):7–26. https://doi.org/10.1007/s42330-019-00070-w
Taber K (1989) Energy—By many other names. Sch Sci Rev 70(252):57–62
Tatar N, Yıldız E, Akpınar E, Ergin Ö (2009) A study on developing a self-efficacy scale towards science and technology. Eurasian J Educ Res 36:263–280
Tiedemann J (2002) Teachers' gender stereotypes as determinants of teacher perceptions in elementary school mathematics. Educ Stud Math 50(1):49–62
Trautner HM, Ruble DN, Cyphers L, Kirsten B, Behrendt R, Hartmann P (2005) Rigidity and flexibility of gender stereotypes in childhood: developmental or differential? Infant Child Dev 14:365–381
UNESCO (2023). https://www.unwomen.org/en/news-stories/op-ed/2023/03/op-ed-technology-and-gender-equality-bringing-women-and-girls-to-the-centre-of-innovation
Van den Broeck L, Demanet J, Van Houtte M (2020) The forgotten role of teachers in students' educational aspirations. School composition effects and the buffering capacity of teachers' expectations culture. Teach Teach Educ 90:103015
Webb-Williams J (2018) Science self-efficacy in the primary classroom: using mixed methods to investigate sources of self-efficacy. Res Sci Educ 48:939–961
Zimmerman BJ (2008) Investigating self-regulation and motivation: historical background, methodological developments, and future prospects. Am Educ Res J 45(1):166–183
Zysberg L, Berry DM (2005) Gender and students' vocational choices in entering the field of nursing. Nurs Outlook 53(4):193–198

Learning Physics

The Wave Model of Light According to the Interpretation of Diffraction and Spectroscopy Phenomena by Upper Secondary School Students

Danilo Catena, Marisa Michelini, Lorenzo Gianni Santi, and Italo Testa

Abstract This study analyzes the interpretation of diffraction and optical spectroscopy phenomena by upper secondary school students based on their use of the basic features of the wave model of light. Two teaching interventions were implemented with a sample of 60 students attending the last year of the high school curriculum, according to a subdivision into two groups: the first one ($N = 34$) experienced only a path about optical spectroscopy, while the second one ($N = 26$) experienced a path about light diffraction too. Their conceptual learning and their use of the light model was investigated through a post-test with open-ended questions. The results were compared with a similar pretest or with a control group. The results show how the specific use of the wave model in the interpretation of diffraction phenomena can influence and hinder conceptual learning relating to spectroscopic phenomena. Implications for teaching are briefly discussed.

1 Introduction

This study represents a small part of a larger study concerning the development of teaching proposals to improve conceptual learning about the wave model of light for high school students. The theoretical framework adopted for this elaboration is the Model of Educational Reconstruction (Duit et al. 2012) with Design-Based Research methods (Collins et al. 2012). By adopting this framework, the literature on students' learning difficulties and on the innovative teaching proposals implemented to try to overcome these difficulties were first analyzed. In doing so, we referred to the wider context of electromagnetic waves, in order to not restrict the analysis to the

D. Catena (✉) · M. Michelini · L. G. Santi
Department of Mathematics, Physics and Computer Science, University of Udine, Udine, Italy
e-mail: catena.danilo@spes.uniud.it

I. Testa
Department of Physics, University Federico II, Napoli, Italy

I. Testa et al. (eds.), *Connecting Physics Education Research and Practice*,
Challenges in Physics Education, https://doi.org/10.1007/978-3-031-86609-8_3

specific context of light. This approach, following the idea of Ambrose and colleagues (Ambrose et al. 1999), aimed at broadening and deepening the results coming from literature, which mostly focused on the light context. In the next section, we reported the more relevant results representing the starting part of our research about the students' understanding of the wave model characterizing electromagnetic waves, in order to point out how the research question we investigated in this study will be framed in this wider context.

2 Background

The main learning difficulties regarding electromagnetic waves—which can be declined in the context of light in different ways—can be classified as follows:

1. Difficulty in understanding the interdependence between electric and magnetic fields in the propagation of an electromagnetic wave, namely, that the electric field cannot be zero in a region of space if the magnetic field is not zero—and vice versa—and that the two fields have to satisfy a specific relationship. In this case, the link with the polarization of light is evident (Ambrose et al. 1999). Ambrose and colleagues also suggested that this difficulty could be caused by difficulties in the correct interpretation of the different kinds of formal representations of an electromagnetic wave (e.g., the wave profile representation reporting the oscillating fields over time or space).
2. Difficulty in identifying the mechanisms of interaction of the electromagnetic waves with matter, according to the Lorentz force (Rettob et al. 2021).
3. Difficulty in identifying the sources of an electromagnetic wave, and in relating them with the different kinds of light sources (Buongiorno and Michelini 2018).

Here we do not focus on general difficulties about the definition of physical quantities characterizing the wave model (e.g., wavelength).

In the literature there is a lack of teaching proposals aimed at overcoming these difficulties, and the main ones basically retrace those individuated by the main textbooks of electromagnetic (Griffiths 2013). On the other hand, there are some tutorials (Ambrose et al. 1999) that exploit the conceptual content of Maxwell's equations to overcome these learning difficulties (e.g., interdependence between electric and magnetic fields). Emphasis is also placed on the Hertz experiment and its historical role in providing early evidence of the propagation of electromagnetic waves, but also on the role it may play in the epistemological reconstruction of the conceptual knots (https://phet.colorado.edu/it/). Finally, interactive simulations and applications are mostly related to optical phenomena or to the simple qualitative illustration of the propagation of an electromagnetic wave (Wosilait et al. 1999). On the other hand, many didactic proposals have been implemented in literature regarding the phenomena of light-matter interaction (e.g., diffraction) (Michelini et al. 2007). In the next section, we specified how we suggest linking these two research fields.

3 Research Question

Within the theoretical framework we chose, the elaboration of teaching proposals about electromagnetic waves requires the analysis of the conceptual referent which the interpretation of the electromagnetic spectrum is based on. Therefore, it is necessary to promote the ability in recognizing the common nature of entities that refer to different models (e.g., wave, radiation, ray) and in interpreting the same phenomena adopting different models. To promote this ability, it is necessary to master the main characteristics each model refers to. In this context, light represents the most suitable portion of the electromagnetic spectrum which to focus on, due to the richness of observable daily experiences, and above all the coexistence of the same model for different phenomena or even the coexistence of different models for the same phenomenon. Thus, we decided to investigate the interpretative application of the wave model of light in the analysis of diffraction and optical spectroscopy phenomena, as provided by students. This study refers to the implementation with high school students of educational proposals about these phenomena, developed by the research unit in physics education of the University of Udine, whose rationale is briefly illustrated in the next section.

The research question that guides this study is therefore related to a very specific aspect and can be formulated as follows: Does the use of a specific model of light in interpreting a phenomenon (namely, diffraction) influence the interpretation of other phenomena (namely, spectroscopy) relating to the same physical entity (the light)? In particular, we aimed at pointing out the results about the wave model.

4 The Rationale of the Educational Proposals

The first educational proposal we implemented is related to optical diffraction (Ivanjek et al. 2020). To clarify the meaning of the results we obtained, we briefly illustrate the steps that characterize the rationale of the path:

1. The characteristic quantities related to the phenomenon of the passage of light through a large slit are identified.
2. We empirically observed what happens when the slit narrows, verifying the new relationships between the characteristic quantities of the phenomenon.
3. An experiment is carried out with the Lucegrafo device (Ivanjek et al. 2020), measuring the distribution of light intensity as the width of the slit varies, in order to identify the law of the diffraction phenomenon from a single slit.
4. The phenomenon is interpreted according to the Huygens-Fresnel principle and the bisection method in the Fraunhofer limit.

The second educational proposal is related to optical spectroscopy, characterized by the following fundamental steps:

1. Identification of the various kinds of light sources, with their operating principles and technologies.
2. Characterization of the emitted light in terms of color and intensity, with collection of emission spectra from different sources.
3. Relationship between emitted light and emission process.
4. Analysis of the structure of matter in terms of energy levels and description of atoms.
5. Transition from Kirchoff's and Bunsen's spectra to Balmer's and Rydberg's laws.
6. Analysis of the quantum nature of light and introduction of the concept of photon. Following these proposals, the study was carried out as described in the next section.

5 Methods

The research sample consisted of 60 students attending the last year of a scientific lyceum in Treviso. The sample is divided into two groups: the first one (consisting of 34 students) experienced the proposal about spectroscopy pathway, while the second group (consisting of 26 students) experienced both educational paths. Each of the two teaching interventions lasted 6 h. At the moment of the implementation, the participants were not yet introduced to these topics by their teachers. All the interventions were carried out by a researcher, and the one about diffraction was implemented before that about spectroscopy. There were significant differences about how the students learned about the wave model of light: in the control group the characteristics of this model were provided at the beginning of the intervention, while in the experimental group the model was built up as a product of the interpretation process about the observed phenomena. Finally, the control group was in a different school than the experimental group, but the researcher who carried out the teaching intervention was the same. The traditional approach experimented by the control group included the experiment about hair diffraction with laser light, that was carried out after illustrating the main characteristics of the wave model of light.

The students' conceptual learning was measured through open-ended post-tests, then the answers were analyzed according to categories defined ex-post in order to be mutually exclusive. The categorization method is iterative. The results about conceptual learning are obtained by analyzing the meaning of the categories obtained and comparing them with the corresponding ones coming from a control group of the same size (for the diffraction path) following a traditional approach, or with the corresponding results of a pretest (for the spectroscopy path). In the first case, the comparison with the control group was performed by means of a chi-square test. In the next section we reported the results we used to answer the research question.

6 Results

In Table 1 the results related to the items about the diffraction phenomenon are reported, also considering the ones coming from the control group. Our analysis reveals 3 categories for both questions, whose operational description is made explicit. The first category includes the answers in which the explanation of the diffraction pattern focuses on the phase difference, without referring to the differences in the optical path. The second category includes the answers stating the dependence of the position of the diffraction maxima on the difference in the optical path between the incident rays at the various points of the slit, sometimes providing simple examples; finally, the third category includes the answers that explicitly report the dependence of the interference condition (constructive or destructive) on the relationship between the incident wavelength and the optical path difference of the rays incident on the slit.

A similar pattern emerges from the analysis of the second question: the first category represents the answers simply asserting that the constructive or destructive interference is given by the fact that the waves composing the incoming radiation are in phase or out of phase. The second category includes those answers highlighting, through simple examples, the dependence of the diffraction pattern (i.e., the positions of maximum and minimum intensity) on the phase difference between the component waves. Finally, the third category expresses the relationship between the diffraction pattern and the initial phase difference between the secondary sources of spherical waves (according to the Huygens Fresnel's principle). In both cases, it is evident that the three categories are ordered according to the completeness of the answer. The chi-square test is significant for both the first item ($p = 0.007$) and the second one ($p = 0.044$).

The items about spectroscopy mostly regarded the role of the dispersive elements that produce the spectra. In Table 2 the emerging response categories for two questions are reported, which were selected and reformulated by previous tools (Tucci 1998). In the first case, we asked which characteristics of a continuous spectrum

Table 1 Items about the diffraction phenomenon. The last two columns represent the frequencies for each category of the two groups: I = Innovative approach ($N = 26$), T = Traditional approach ($N = 27$)

Items	Categories	I	T
How would you use the concept of optical path to explain the diffraction pattern formation	1. Optical path = Phase difference	7	3
	2. Optical path influences diffraction pattern	8	20
	3. Relationship between wavelength and optical path	11	4
How would you use the concept of optical path to explain the diffraction pattern formation	1. Optical path = Phase difference	11	8
	2. Optical path influences diffraction pattern	6	15
	3. Relationship between wavelength and optical path	9	4

would change after removing the prism in the experimental device that provided it. 4 categories emerged: (i) the spectrum will not change; (ii) the linear dimensions of the spectrum will change; (iii) the spectrum will no longer be continuous; (iv) A white continuous region will appear. In a similar question concerning the replacement of the prism with a diffraction grating, the same categories emerged, with two further ones: (i) the observed spectrum is equivalent to the original one as if it were observed in the mirror (inconsistent answer); (ii) you will observe a white central band and side stripes of the various colors of the rainbow. In particular, it is interesting to observe that there is an overlap in the two questions between the answers of students who believe that the spectrum will not change and those who believe that it will become discrete.

A similar question is posed about discrete emission spectra, asking to indicate one—or more—ways in which the spectral shape can be modified. The answers are shown in Table 2, and since a high variability has been observed, the categories are

Table 2 Items about the spectroscopy phenomenon. The last two columns represent the frequencies for each category of the two groups: I = Innovative approach ($N = 26$), C = Control group ($N = 34$)

Items	Categories	I	C
Consider a continuous spectrum How will the spectrum change removing the prism	1. It does not change	3	9
	2. Not continuous anymore	6	3
	3. The linear dimension changes	2	5
	4. White region	15	17
Consider a continuous spectrum How will the spectrum change replacing the prism with a diffraction grating	1. It does not change	4	11
	2. Not continuous anymore	10	4
	3. The linear dimension changes	7	2
	4. White region	3	7
	5. Image in the mirror	0	4
	6. White central region and side rainbow bands	2	6
Consider a discrete spectrum How is it possible to change its shape *Categories not mutually exclusive	1. Changing the size of the slit	27	16
	2. Changing the source	9	14
	3. Using lenses	4	10
	4. Changing the alignment of the grating	1	7
	5. Changing the shape of the slit	9	12
	6. Changing the source-grating distance	8	12
Which model of model does justify the observation of an emission spectrum	1. Ray model	7	6
	2. Photon model	4	13
	3. Wave model	14	12
	4. Mix of models	1	3

not mutually exclusive. Here we just want to emphasize that the most populated category refers to the slit, and not to the source.

Finally, in order to answer our research question and to link the information coming from the previous questions, a general item was asked about the light model that allows us to justify the observation of a generic emission spectrum. The results show considerable variability, so that the responses can be classified into 4 categories, as shown in Table 2. The wave model is preponderant compared to the ray and photon models, as it also occurs about the identification of the characteristic quantities of the light sources. On the other hand, some answers can be considered part of the category of "mix of models", as descriptive and interpretative aspects from different models coexist.

Two different columns are reported in Table 2, in order to compare the frequencies for each category of two groups of students: the former (I) experienced the innovative teaching approach about diffraction phenomena, while the latter was not introduced to diffraction at all. Considering the description of the categories we provided above, the results clearly show that the construction of the wave model of light as an interpretative key of diffraction phenomena has influenced the interpretation of optical spectroscopy phenomena.

7 Discussion

Regarding diffraction, the significance of chi-square tests shows how students who have experienced the innovative path make greater use of the scientific model that interprets the phenomenon of optical diffraction based on the conceptual referents used in the path. In the case of spectroscopy, on the other hand, the clear overlap (for more than a third of the sample) that emerges among those who believe that the removal of the prism or its replacement with a diffraction grating transform a continuous spectrum into a discrete one. The answers of these students are therefore consistent with the attribution of a dispersive character to the slit that produces the spectrum. This overlap is almost total for students who have experienced the diffraction proposal before that about spectroscopy. It emerges even more regarding the sample of students who consider the variation in the size of the slit as the only feature who is able to modify the shape of a spectrum, without any reference to the characteristics of the source. Finally, the prevailing light model to justify the formation of a spectrum through a typical experimental setup (light source, mask with slit, prism and screen) turns out to be the wave model. From the students' answers, it is clear that this may be due to the fact that it characterizes the source in a complete way, providing physical quantities (e.g., wavelength, light intensity) that uniquely describe it. Moreover, it overlaps less with the other two models, while there are elements that in the students' interpretation the ray model and the photon model are sharing. These common aspects arise confusion between the two models in students' reasoning: for example, students consider photons as particles of the same nature as those that constitute the "ray" entity of the geometric optics.

8 Conclusions

Although the analysis of other items show a good mastery acquired by students about the descriptive knowledge of optical spectroscopy phenomena, the results we reported above suggested that the interpretation of these phenomena is strongly influenced by the wave model of light, in particular if it was built up to interpret other phenomenon such as diffraction. For instance, the students who experimented the teaching proposal about diffraction before that about optical spectroscopy, tend to attribute a dispersive role to the slit that produces the discrete spectrum, as can be seen from the answers provided to different items. This conceptual bias suggests that dealing with diffraction and polarization phenomena in optics both through a wave model of light and through a photon model as a quantum object (which was used as conceptual referent for the education proposal about optical spectroscopy), while at the same time overcoming the narrow separation between classical and modern physics in the curriculum.

References

Ambrose BS, Heron PRL, Vokos S, McDermott LC (1999) Am J Phys 67:891–898
Buongiorno D, Michelini M (2018) J Phys Conf Ser 1287
Collins A, Joseph D, Bielaczyc K (2012) J Learn Sci 13:15
Duit R, Gropengieber H, Kattmann U, Komorek M, Parchmann I (2012) A framework for improving teaching and learning science 13
Griffiths DJ (2013) Introduction to electrodynamics. Pearson Education, Boston
https://phet.colorado.edu/it/
Ivanjek L, Shaffer P, Planinic G, McDermott L (2020) Phys Rev Phys Ed Res 16:010102
Michelini M, Stefanel A, Santi L (2007) The impact of educational research on physics teacher preparation
Rettob RJ, Poluakan C, Tulandi DA, Mongan SW, Polii J (2021) J Phys Conf Ser 1968
Tucci P (1998) La fisica nella scuola 5
Wosilait K, Heron PRL, Shaffer PS, McDermott LC, McDermott LC (1999) Am J Phys 67

Enhancing Integrated Learning of Physics and Mathematics in Secondary School

Stefania Lippiello, Marta Carli, and Ornella Pantano

Abstract The interplay between mathematics and physics is a pivotal focus of physics education research. Our research into the challenges students encounter in using derivatives, integrals, and vectors in undergraduate physics courses has recently been extended to secondary schools, with the aim of developing research-based teaching activities. In this research, we report our work with a teacher learning community aimed at creating such activities. The Test of Calculus and Vectors in Mathematics and Physics, which we originally developed for university students and then adapted for secondary school students, was used both as a pre/post-survey tool and as a discussion catalyst. Teaching-learning activities were co-designed with the teachers. Specifically, we proposed using Uhden's modeling cycle, augmented by a meaningful use of various representational forms, as a model for activity design. The project involved 260 students and 9 teachers from 4 schools. In addition to the test, we collected teachers' logbooks and conducted interviews to gain insights into classroom activity progression. Our findings underscore the potential of an integrated mathematics-physics approach but also highlight the crucial role of contextual factors and the specific ways collaboration is enacted for its efficacy. Professional development opportunities that foster authentic collaboration are therefore needed.

1 Introduction

The relationship between mathematics and physics is profound and intricate. Physics relies heavily on mathematical concepts and tools to model, analyze, and understand the natural world. Understanding this relationship is essential for developing effective

S. Lippiello · M. Carli · O. Pantano
Dipartimento di Fisica e Astronomia, Università di Padova, Padova, Italy
e-mail: marta.carli.1@unipd.it

O. Pantano
e-mail: ornella.pantano@unipd.it

S. Lippiello (✉)
Liceo Scientifico Jacopo da Ponte, Bassano del Grappa, Italy
e-mail: stefania.lippiello@phd.unipd.it

I. Testa et al. (eds.), *Connecting Physics Education Research and Practice*, Challenges in Physics Education, https://doi.org/10.1007/978-3-031-86609-8_4

teaching methods and promoting deeper learning in both mathematics and physics. In this work, we focused on secondary school to study *how participation in a teacher learning community centered on the structural role of mathematics in physics can impact teachers' practices and its effect on students.* To investigate this research question, we organized a training course for in-service teachers structured as a teacher learning community, with the aim of improving the integrated teaching of mathematics and physics. The specific tools utilized to answer the research question were a pre/post test, used to monitor the students' improvements, and individual interviews with participating teachers to gain insights into the enactment of the activities in the classroom and the process of teacher collaboration.

2 Theoretical Background

Contemporary research has moved away from traditional perspectives that confined mathematics to a mere calculating tool and, on the other hand, considered physics as nothing more than a mere setting for applying these calculations. Mathematics serves as both a language and a mode of reasoning for physics (Redish and Kuo 2015; Kuo et al. 2013), yet their relationship can be explored even more profoundly. Uhden et al. (2012) emphasized the structural role of mathematics in physics, connected to a way of logical reasoning, as opposed to the technical role related to manipulations of variables. They support their argument by citing historical instances when the development of one field was inseparable from the other (see also (Williams et al. 2016; Karam et al. 2019)). Indeed, even seemingly basic physics concepts like velocity—a rate—and acceleration—a rate of a rate—are not "just physical but already mathematical concepts in themselves" (Uhden et al. 2012, p. 493). This has implications for physics educations, since learning to proficiently use mathematics in physics requires developing skills corresponding to both the structural and the technical role. Starting from previous models (Bing and Redish 2009; Blum and Borromeo Ferri 2009; Pietrocola 2008; Blum and Leiss 2005), Uhden et al. (2012) propose a refined model for physical-mathematical reasoning and problem solving that explicitly distinguishes between structural and technical skills and effectively incorporates mathematics into physics education. The cycle consists of a series of steps that, starting from the real world, involve simplification, mathematization, interpretation, technical-mathematical operations and, finally, validation of the model thus created to describe a real-world phenomenon or problem. The steps of this model where the interplay of mathematics and physics becomes more subtle are mathematization (formalizing a physical problem with gradually increasing degrees of mathematization) and interpretation (physical meaning is deduced from equations, identifying special cases or making physical predictions from the formalism). In these phases, particular attention is given to how a problem can be represented, through graphs, words, numbers, or equations. The ability to switch between these different languages or representations, known as representational fluency, is a key indicator of expertise in physics (Van Heuvelen 1991; Etkina et al. 2006).

Students' awareness of such a deep interrelation between mathematics and physics cannot be taken for granted. Proficiency in solving end-of-textbook physics problems does not necessarily guarantee conceptual understanding of mathematical tools in the context of physics. Even in the absence of such understanding, students may be able to solve quantitative physics problems using a "plug-and-chug" approach (Tuminaro and Redish 2007). Conversely, ability in the use of mathematical tools does not assure success in physics (Redish 2006; Greca and Moreira 2002). Redish (2006) points out that the use of mathematics in physics is "semantically different" from its use in a purely mathematical context. Therefore, the teaching of mathematics and physics should surpass the "mathematics as prerequisite illusion" (Karam et al. 2019) in favor of an intrinsically integrated approach since the school level.

To promote instruction that values and leverages the structural relationship between mathematics and physics, we propose to use Uhden's model as a design tool. This approach was used to ground and shape the activities of the teacher training course.

In order to bridge the gap between research and practice, an effective approach is that of teacher communities (Vangrieken et al. 2017; Levy et al. 2021; Carli and Pantano 2023), i.e., groups of teachers who meet regularly over an extended period of time to collectively advance their expertise on a professional topic of common interest. In our group, we have framed in-service teacher training into teacher communities since 2018, starting with the project *CoLLabora—A community of learners on laboratory work* (Carli and Pantano 2023, 2021) and applying the same framework to our in-service teacher training programs consistently (Carli et al. 2003, 2025) since then the relationship between teachers and physics education researchers is modeled as "authentic collaboration" (Couso 2008, 2015), with teachers and researchers collaborating to design and implement teaching sequences. In this project, we have further emphasized the topic of collaboration by also encouraging participation of teachers from the same school, as suggested for example by Desimone (Desimone 2009). Deep collaboration between teachers—where beliefs and practices are openly discussed—improves instructional efficacy and contributes positively to school-wide change. Structured collaboration increases motivation, reduces individual workloads, supports the implementation of student-centered strategies and ensuring alignment between the written curriculum and instructional practices, and can positively influence student performance (Vangrieken et al. 2017).

3 Methods

To answer our research question and evaluate the effectiveness of the training course, we studied how the experience of a group of teachers in a learning community centered on the relationship with mathematics and physics impacted their practices and the effect of the co-developed activities on students. Specifically, we used three instruments: a modified version of an instrument we have previously developed, the Test of Calculus and Vectors in Mathematics and Physics (TCV-MP) (Carli et al.

2020); a logbook to record what teachers did in class; and individual interviews with the teachers at the end to get further feedback on the class process and the project in general.

The TCV-MP is a multiple-choice test that we developed in 2018 for first-year students at our university. It was aimed at facilitating students' self-assessment regarding the use of derivatives, integrals and vectors at the beginning of their Physics 1 courses. Built upon both an examination of the literature and an analysis of common students' mistakes, the test consists of 34 questions: 17 items in mathematics (6 on derivatives, 3 on integrals, and 8 on vectors) and an equivalent set of 17 items in physics (mechanics, focusing on kinematics or point dynamics). The selection of mathematical tools and physics topics was based on their significance for introductory physics courses. The items involve various representational formats: natural language, graphs, numerical data, and formal mathematical expressions and equations. The full breakdown of the test structure, inclusive of the mathematics and physics topics and the representations used in each item, is reported in (Carli et al. 2020). In a previous project, we adapted the test to make it suitable for high school students from both fourth- and fifth-year students (grades 12 and 13) (Lippiello et al. 2022). Since the formal definition of derivatives and integrals is not given until the half of the fifth year, we modified the language of the calculus items to make them understandable to the students. Specifically, the derivative was described as "the slope of the tangent line to the graph of a function at a point" and the definite integral was described as "the area under the curve". No modifications were necessary for the items on vectors, as these are typically covered in the early years of secondary school. For clarity, an example of a pair of items from the high-school version of the test is provided in Fig. 1.

In this project, the test was used as a pre/post tool; in between, the teachers were encouraged to conduct an experimentation in the classroom based on the ideas and activities discussed during the training course. We collected pre/post data from 14 classrooms (13 fifth-grade, 1 fourth-grade; 260 students). Logbooks were used to document the experimentation in the classroom. Teachers were asked to write notes and comments about the activities and any other useful observations they made during the experimentation. The logbook was compiled online on the course platform.

Semi-structured interviews were conducted with individual teachers just after the conclusion of the course, to gain insights into the processes and factors that influenced the experimentation, teachers' perception of the applicability of Uhden's model for the design of instructional activities, and the process of collaboration. As the interviews were primarily focused on the actual enactment of the course in the classroom, we only interviewed the teachers who conducted an experimentation (9 of the 12 participating teachers). The interview protocol was developed based on the following categories, each exemplified by a sample question: *perceived professional development* (e.g., Is there a professional area in which you feel you have strengthened your skills?); *collaboration* (e.g., Were you able to collaborate with other teachers, either those who attended the course or colleagues from your school?); *relationship between mathematics and physics* (e.g., How did you view the relationship between mathematics and physics before, and how do you see it now? Has your perspective changed in any way?); *classroom implementation and effects on students* (e.g., Did

Item 3M. The figure below shows the graph of a function f defined on $\mathbb{R}$.

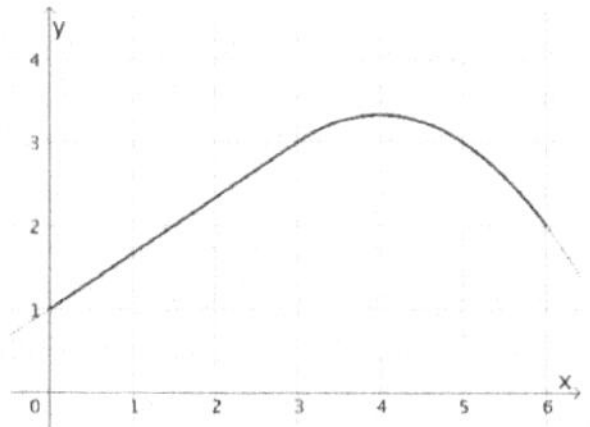

Let g be the function that associates, with each x, the slope of the tangent line to the graph of f at the point $(x, f(x))$. What is the value of $g(3)$?

(A) 2/3

(B) 3

(C) 3/2

(D) 1

(E) 1/3

Item 3P. The figure below shows the position-versus-time graph of an object.

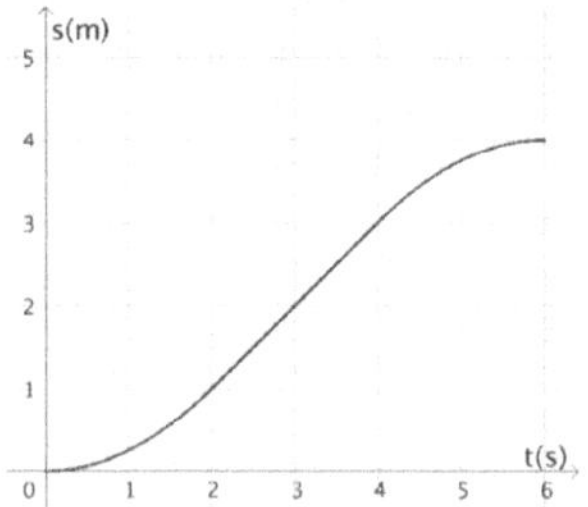

What is the object's velocity at t=3s?

(A) 1.0 m/s

(B) 2.0 m/s

(C) 1.5 m/s

(D) 0.67 m/s

(E) 0.5 m/s

Fig. 1 A pair of items from the revised test: 3M is set in the mathematical context, while 3P is its isomorphic counterpart in the physical context. Note the choice of language, which avoids the use of the term derivative

you notice any effects on the students? What was the most noticeable effect? Did you receive feedback from students or colleagues?); *additional comments and future perspectives* (e.g., Are you interested in continuing training on these topics?).

4 Implementation

The experimentation took place in the school year 2022/23. The learning community comprised 12 in-service mathematics and physics teachers from 5 high schools in the Veneto region in Italy. All schools were "scientific" high schools ("Liceo Scientifico"), i.e., schools with a dominant scientific component (specifically, students get 5 years of compulsory physics teaching). Table 1 summarizes the composition of the group in terms of schools, classrooms, and teachers, specifying the subjects they taught in the classrooms involved. To avoid any reference to the specific institutions and teachers involved, the schools were labeled A-D and the teachers T1-T9. One of the teachers (T9 from School D) started the experimentation but did not conclude it. However, we decided to include him in the interview dataset.

We decided to specify the subject(s) taught in each class (mathematics, physics, or both) because this detail turned out to be relevant in the implementation. For example, in school B there is a policy according to which teachers normally teach only one of the two disciplines (either mathematics or physics), while another colleague teaches the other discipline in the same classroom.

Table 1 Participant schools, classrooms, and teachers. Classes marked with a star were involved in the experimentation through sharing of teaching materials between colleagues in the same school, but their teacher did not take part in the training program

School	ID of classes	Teachers (and taught subjects)
A	5A, 5B, 5C	T1, T2, T3 (teaching both mathematics and physics)
B	5A, 5B*, 5C*, 5D*, 5E, 5F*, 5G, 5H, 4A	T4 (mathematics in 5A and 4A), T5 (physics in 5E and 5G), T6 (physics in 5H)
C	5A	T7 (mathematics and physics)
D	5A, *5B*(pre-test only)	T8, *T9* (teaching both mathematics and physics)

Table 2 Schedule and content of the program

Date	Activity
28/10/2022 (3h)	Presentation of the TCV-MP and reflection on its use in the classroom
11/11/2022 (3h)	Analysis of the needs of different contexts; discussion of Uhden's model and design of a sample sequence
November–December	Pre-testing of students. Data analysis and sharing with teachers
10/02/2023 (3h)	Discussion of the results of the pre-test and collaborative design of teaching activities based on test results and emergent needs
Febryary–April	Classroom implementation of activities
May	Post-testing of students. Data analysis and sharing with teachers
June–July	Individual interviews

The schedule and topics of the program are reported in Table 2. During the first session, we presented the TCV-MP instrument and we had the teachers try to solve it themselves. Then, we collectively discussed it to ensure they grasped its characteristics and objectives. This enabled the whole group to proceed with understanding how to integrate the project into their specific contexts by analyzing their unique needs. Subsequently, we introduced Uhden's model and we showcased an example of a teaching sequence on motion designed using the model, with the main mathematical tool involved being the derivative. As an exercise, the teachers were asked to outline a similar teaching sequence using the integral as the main mathematical tool.

After the pre-test implementation, we discussed the results with the teachers to highlight common difficulties and to identify specific needs. Participants then collaboratively designed the experimentation, working with colleagues from the same school and/or with other participants with similar specific interests or classrooms.

As suggested above, we proposed to design the activities based on Uhden's modeling cycle (Uhden et al. 2012). This approach constitutes a novelty compared to usual activities, and was aimed at making the relationship between mathematics

Mathematization → **Interpretation** →

Verbal expression	Variables	Formula	Implications
A population of bacteria grows at a constant rate of 1800 bacteria per hour.	N = number of bacteria t = time (hours)	dN/dt = 1800	• rate = slope of the tangent line to the graph of the function expressing the number of bacteria as time progresses • rate = derivative of the function expressing the number of bacteria as time passes • the slope of the tangent line at each instant of time is equal to 1800 • the function associating the rate at each instant of time (the derivative function) is a constant function • 1800 > 0 => the function expressing the number of bacteria increases

Fig. 2 An example of activity focused on the processes of mathematization and interpretation

and physics explicit and distinguish technical and structural skills. For example, we proposed the following teaching sequence on derivatives:

- *Simplification*: from the real world, a ball rolls up the ramp and then back down, to the motion of a material point.
- *Mathematization*: representation of the situation with graphs; formalization through the concepts of average and instantaneous velocity and subsequent levels of formal mathematization (derivatives).
- *Interpretation*: from laws, equations, and graphs, what can be deduced from a physical point of view?
- *Mathematical operations (technical aspects)*: the slope of a tangent line as the derivative, a function and its derivative function.
- *Validation*: back to the real world.

Each step of the process was then developed with specific activities. For example, in Fig. 2 we provide an excerpt from a worksheet focused on the processes of mathematization and interpretation. Starting from a sentence in natural language, students were asked to "translate" it into mathematical language in gradually more formal steps and finally interpret everything that could be deduced from its formalization. The same exercise was proposed through different contexts, both in physics and outside physics.

Additionally, we placed emphasis on the use of different representations. This was supported, for example, with a GeoGebra book, an excerpt of which is shown in Fig. 3. The book contains interactive exercises similar to the ones in the TCV-MP, through which the students can work independently in small groups in the classroom or individually at home.

Strategies to maximize the effectiveness of these activities, such as group work, were modeled throughout the course and explicitly discussed. Flexibility was provided to teachers in determining how to implement these activities in their classrooms, allowing them to adapt the approach to their specific needs. Specifically, teachers

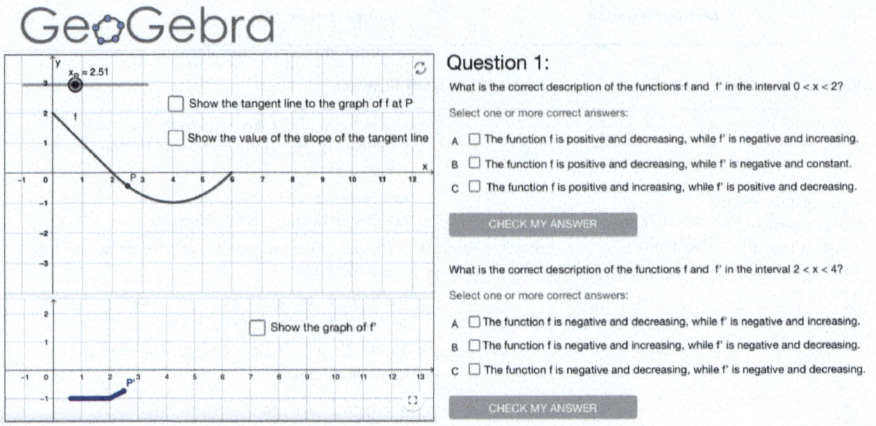

Fig. 3 A screenshot from the GeoGebra book displaying an interactive exercise example. The text in the image has been translated from Italian into English

T1, T2, and T3 chose to implement the full Uhden cycle, while the others focused primarily on the processes of mathematization and interpretation. As revealed by logbooks and interviews, most teachers decided to focus on derivatives and integrals rather than on vectors. On average, teachers allocated six teaching periods, each lasting 50 min to one hour, to these activities.

5 Results

Table 3 shows descriptive statistics both for the test as a whole and by subject (considering the 17 maths and physics items separately, normalized to a 10-point scale), considering the whole sample (all students from all schools).

Figure 4 helps visualizing and comparing pre- and post-test results. A Wilcoxon signed-rank test showed a statistically significant difference between the pre and post test scores with a large effect size ($W = 1481.5$, $p < 0.001$, $r_b = 0.90$). The difference between the mathematics and physics part was not statistically significant

Table 3 Descriptive statistics

	Pre		Post	
	Median	Dev. St.	Median	Dev. St.
Test score	5.29	1.89	7.06	1.78
Math score	5.88	2.11	7.36	1.86
Phys score	5.29	2.02	7.06	1.99

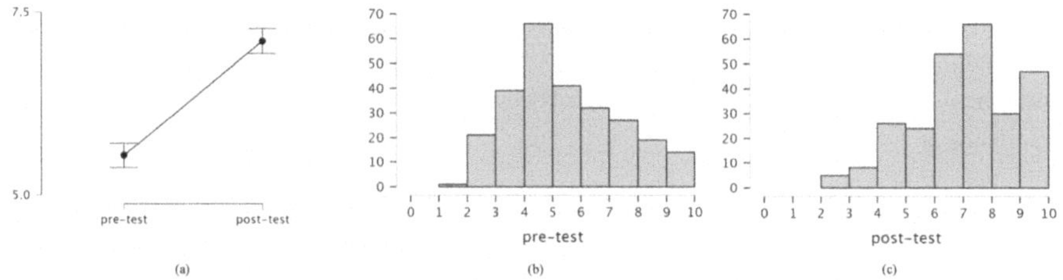

Fig. 4 **a** Descriptive plot showing the comparison between the pre-test and post-test scores. **b** Distribution plot for the pre-test scores. Absolute frequency is reported on the y-axis. **c** Distribution plot for the post-test scores

Table 4 Descriptive statistics per mathematical tool employed (derivatives, integrals, or vectors)

	Pre				Post			
	Math		Phys		Math		Phys	
	Median	Dev. St.	Median	Dev. St.	Median	Dev. St.	Median	Dev. St.
Derivatives	5.00	2.65	5.01	2.62	6.67	2.27	6.67	2.60
Integrals	6.67	3.52	5.00	3.38	10.00	2.71	10.00	3.17
Vectors	6.25	2.28	6.25	2.18	6.25	2.27	7.50	2.18

(pre math-phys: $p = 0.177$; post math-phys: $p = 0.097$), and test scores in the two domains were correlated (pre: $r = 0.689$, $p < 0.001$, post: $r = 0.713$, $p < 0.001$).

The breakdown of results according to the mathematical tool employed (derivatives, integrals, or vectors) are presented in Table 4 (normalized to a scale of 10). An improvement can be observed across all areas between the pre-test and post-test, both in mathematical and physical contexts. In the pre-test, the weakest area was related to derivatives. Notably, the scores for vectors in the mathematical context did not show any variation. This outcome can be attributed to the lack of targeted instruction on this topic by the teachers, as evidenced by an analysis of the logbooks and interviews, which revealed a stronger focus on derivatives and integrals.

Class-level analysis provided valuable insights for teachers, offering a comprehensive overview of their class's performance. From a research perspective, collecting additional data on measurement covariates would be required to analyze and compare differences across classes effectively. Thus, we now focus on and contrast the two schools with the highest number of participating classes and teachers (School A with $N = 55$ and School B with $N = 153$). There is no significant difference in the pre-test scores ($t = 0.518$, $p = 0.606$) suggesting that the two schools started from a comparable level. In contrast, in the post-test, the difference between the two schools becomes significant in favor of School A ($t = 3.407$, $p = 0.001$, $d = 0.459$) with a medium effect size.

The interviews with teachers from the two schools are useful to better interpret these results. In School A, the three teachers work and experiment in parallel in a collaborative manner. Each of them teaches both mathematics and physics in the

Table 5 Number of teachers who mentioned each category in the interviews, with example quotations

Category	N teachers	Examples
Importance of collaboration	9	"I was especially enriched by the face-to-face meetings with colleagues from other schools and sharing some of the paths with them. This enriched me a lot and I enjoyed it: the opportunity to interact with colleagues from other schools [...], the collaboration both with you at the university and with colleagues [from my school]." (T3)
Difficulties with colleagues at the school	9	"I had hoped to be able to collaborate more with my colleague, since we both took the course, but we got out of sync as the lectures went on and it was increasingly difficult to work together." (T8)
Difficulties with time management	9	"The greatest challenge I faced was time constraints; I would have needed more time." (T3)
Interest in the Uhden model, more time needed	8	"It helped systematize things I was doing without awareness. [...] Now I have to let things settle a bit." (T4); "The steps help make the processes explicit. It is to be deepened in the future." (T3)
Difficulties in relating with students	2	"The students showed resistance, they did not accept the job willingly." (T9)

classes involved, and the students were explicitly introduced to the project. In School B, instead, only one teacher developed the materials and shared them with colleagues (only three of which were directly involved in the course, whereas the other did not participate in the training). Therefore, although this school had more classes participating, there was less involvement of both teachers (not all being equally trained) and students (not all classes were adequately introduced to the project). Furthermore, as mentioned above, in this school the teaching of mathematics and physics is separate (ach teacher teaches only one of the two subjects in a certain classroom).

We analyzed the complete dataset of interviews conducted with the nine teachers who implemented the experimental approach, employing an inductive method to identify recurring themes related to their perspectives on the proposed methodology. The key findings, in terms of the most frequently mentioned topics, are summarized in Table 5. To illustrate how the codes emerged, we provide representative quotations from the interview transcripts in the final column.

All nine interviewed teachers underscored the significance of collaboration. For instance, one teacher remarked, "We always work together, we confide in each other a lot. [...] If there wasn't [the group] I would be lost. Collaboration is crucial, and this, I think, was a concrete collaboration" (T2). Another highlighted the value of collegial interaction, stating, "What I take home is the discussion with other teachers [...]. Not being alone was important. At school, collaboration should be established as a standard practice. What I missed instead was the collaboration with my colleague teaching physics [in the same classroom]" (T4). This latter statement also underscores how collaboration, while valued, can pose challenges when it is absent or insufficiently structured. Furthermore, the teachers expressed a strong belief in the interconnectedness of mathematics and physics. One participant noted, "Doing only physics is probably not possible, and neither is doing only mathematics [...]. You need to create an interaction between the two [subjects], each with its own role" (T2). The activities proposed during the course were generally perceived as practical and immediately applicable in the classroom. For example, one teacher explained, "It was immediately usable with the students, which was fundamental" (T1). Another shared their experience of applying the methods, stating, "I had the opportunity to discuss with other colleagues and with you researchers and to learn innovative methods with which to introduce integrals and manage the class [in terms of group work]. I immediately tried it out with my students and received positive feedback" (T8).

With regard to the logbooks, the thematic analysis was constrained by the varying levels of detail provided by the teachers in their entries. Specifically, most teachers focused on uploading the materials used (e.g., worksheets and exercises) rather than offering descriptions of the methods employed during the lessons. Thus, while the logbooks were valuable for understanding what teachers proposed to their students, the interviews offered a deeper insight into their experiences and the reflections that emerged. For instance, the interviews revealed how the teachers approached their work and the challenges they encountered.

6 Discussion and Conclusions

This study investigated how participation in a teacher learning community centered on the structural role of mathematics in physics can impact teachers' practices and their effect on students. Through the adoption of the framework proposed by Uhden, Karam, Pietrocola, and Pospiech Uhden et al. (2012) and the deliberate use of varied representational forms (Van Heuvelen 1991; Etkina et al. 2006) to design instructional activities, we aimed to promote an integrated approach to mathematics and physics in secondary education. We encouraged teachers to design and implement their own teaching sequence in the classroom and we use different tools to analyze both the outcomes (students' performance, measured using a revised version of a research-based multiple choice test) and the process (through teachers' interviews and logbooks). Our findings indicate an overall gain in test scores, suggesting an improvement in students' use of derivatives, integrals, and vectors. The test not only served as an effective pre/post assessment tool but also fostered discussions among teachers, enabling them to identify specific learning challenges and adapt their strategies. However, the limited duration of the program constrained deeper exploration of the proposed instructional framework, highlighting the need for sustained engagement. Teacher interviews underscored the pivotal role of collaboration in enhancing instructional practices. Working with peers within and across schools facilitated the exchange of ideas and provided support for implementing interdisciplinary approaches. Nevertheless, structural barriers, such as separate teaching roles for mathematics and physics and limited opportunities for coordinated planning, often hampered the full realization of these benefits. Addressing such contextual factors is essential to maximize the impact of teacher learning communities. Moreover, systematically examining contextual variables—such as teacher training backgrounds, classroom dynamics, and institutional policies—will provide deeper insights into the conditions that enhance or hinder the success of integrated teaching approaches. Recent physics education research highlights physical-mathematical modeling not only as a theoretical framework but also as a pedagogy and instructional design tool. For example, Dominguez et al. (2023) described the redesign of separate Physics and Mathematics courses into an integrated Phys-Mat course, emphasizing modeling and representational tools (e.g., drawings, motion graphs, tables, and maps). The course articulates physics and mathematics goals for each topic, focusing on mathematical tools such as functions, derivatives, and integrals. Each cycle begins with a real-world problem requiring physical and mathematical modeling for resolution, with repeated cycles designed to build students' skills. A key finding was that constructing conceptual models before calculations was essential; on the other hand, integrating physics and mathematics fostered conceptual understanding. These results support efforts directed at integrating mathematics and physics and at resolving tensions between conceptual and mathematical approaches (Uhden et al. 2012). As Dominguez et al. (2023) also noted, the challenge remains how to scale this approach to large-classroom settings, such as those for which the TCV-MP was originally developed. The key elements identified in the present study—solid theoretical grounding, careful activity design mindful of the context, and collaboration between mathematics and physics instructors—could be crucial to extending this approach to such environments. Addressing these challenges can enhance the scalability and effectiveness of interdisciplinary education, fostering a more cohesive and impactful STEM learning environment.

Acknowledgements We thank the *Piano Lauree Scientifiche* (PLS—POT 2019–2020, DM 435/2020) of the Italian Ministry of Education, University and Research for financial support, and all the students and teachers who took part in the project.

References

Bing TJ, Redish EF (2009) Epistemic complexity and the journeyman-expert transition. Phys Rev ST Phys Educ Res 5:020108

Blum W, Borromeo Ferri R (2009) Mathematical modelling: can it be taught and learnt? J Math Model Appl 1:45–58 (2009)

Blum W, Leiss D (2009) "Filling up"—the problem of independence-preserving teacher interventions in lessons with demanding modelling tasks. In: Bosch M (eds) Proceedings of 5th congress european research mathematics education. Universitat Ramon Llull, Gerona

Carli M., Leonardi A.M., Ciroi S. & Pantano O.: Teaching physics through astronomy in secondary school: the Asiago Teachers' Network on Astrophysics (under revision)

Carli M, Pantano O (2023) Collaborative physics teachers: enhancing the use of the laboratory through action research in a community of learners. Phys Rev Phys Educ Res 19:020162

Carli M, Lippiello S, Pantano O, Perona M, Tormen G (2020) Testing students ability to use derivatives, integrals, and vectors in a purely mathematical context and in a physical context. Phys Rev Phys Educ Res 16:010111

Carli M, Leonardi AM, Gabelli L, Lippiello S, Lissandron G, Pantano O (2023) Innovating physics teaching through teachers' learning communities and action research. Il Nuovo Cimento 46:209

Carli M, Pantano O (2021) A community of learners on laboratory work. Design and implementation of a teacher training programme. In: Jarosievitz B, Sukosd C (eds) Teaching-learning contemporary physics: from research to practice. Springer

Couso D (2015) Participatory approaches to curriculum design from a design research perspective. In: Psillos D, Kariotoglou P (eds) Iterative design of teaching-learning sequences. Springer, pp 47–71

Couso D (2008) Authentic collaboration: a promising paradigm for physics education reform. Int Newsl Phys Educ 56:7

Desimone LM (2009) Improving impact studies of teachers' professional development: toward better conceptualizations and measures. Educ Res 38(3):181–199

Dominguez A, Becerra J, Quezada-Espinoza M, Zavala G (2023) Integration of physics and mathematics in STEM education: use of modeling. Educ Sci 14:20

Etkina E, Van Heuvelen A, White-Brahmia S, Brookes DT, Gentile M, Murthy S, Rosengrant D, Warren A (2006) Scientific abilities and their assessment. Phys Rev ST Phys Educ Res 2:020103

Greca IM, Moreira MA (2002) Mental, physical, and mathematical models in the teaching and learning of physics. Sci Ed 86:106–121

Karam R, Uhden O, Höttecke D (2019) The "Math as Prerequisite" illusion: historical considerations and implications for physics teaching. In: Pospiech G, Michelini M, Eylon BS (eds) Mathematics in physics education. Springer, Cham, pp 37–52

Kuo E, Hull MM, Gupta A, Elby A (2013) How students blend conceptual and formal mathematical reasoning in solving physics problems. Sci Educ 97:32–57

Levy S, Bagno E, Berger H, Eylon B-S (2021) Professional growth of physics teacher-leaders in a professional learning communities program: the context of inquiry- based laboratories. Int J Sci Math Educ 20:1813

Lippiello S, Carli M, Pantano O (2022) Supporting students' mathematization skills for physics: a test and a learning path for secondary school. In: INTED2022 proceedings, pp 4298–4307

Pietrocola M (2008) Mathematics as structural language of physical thought. In: Vicentini M, Sassi E (eds) Connecting research in physics education with teacher education, vol 2. ICPE—book

Redish EF (2006) Problem solving and the use of math in physics courses, preprint arXiv: preprint-physics/0608268

Redish EF, Kuo E (2015) Language of physics, language of math: disciplinary culture and dynamic epistemology. Sci Educ 24:561–590

Tuminaro J, Redish EF (2007) Elements of a cognitive model of physics problem solving: epistemic games. Phys Rev ST Phys Educ Res 3:020101

Uhden O, Karam R, Pietrocola M, Pospiech G (2012) Modelling mathematical reasoning in physics education. Sci Educ 20:485–506

Van Heuvelen A (1991) Learning to think like a physicist: a review of research-based instructional strategies. Am J Phys 59:891–897

Vangrieken K, Meredith C, Packer T, Kyndt E (2017) Teacher communities as a context for professional development: a systematic review. Teach Teach Educ 61:47–59

Williams J, Roth W-M, Swanson D, Doig B, Groves S, Omuvwie M, Borromeo Ferri R, Mousoulides N (2016) Interdisciplinary mathematics education. A state of the art. In: ICME-13 topical surveys. Springer Open

Educational Proposals

Design of a Teaching/Learning Sequence on Surface Phenomena in Liquids for University Education

Ilaria Grazia, Onofrio Rosario Battaglia, Giulia Termini, and Claudio Fazio

Abstract The physics of surface phenomena in liquids encompasses a wide range of topics, many of which are often underexplored or overlooked in school and university curricula. In this paper, we present and examine elements of a Teaching/Learning Sequence (TLS) on surface phenomena in liquids designed specifically for university students. This TLS was developed by reconstructing the scientific content based on observations gathered from a pilot trial of two previous TLSs on the same topic. In this updated TLS that we call TLS 2.0, students take a more active role in knowledge construction: they are more extensively involved in modeling activities through computer-based simulations compared to the prior trial. This approach aims to assess how greater student participation in modeling activities and simulations can support and enhance scientific learning.

Keywords Physics education · Surface phenomena in liquids · Teaching/learning sequences

1 Introduction

Research in science education has a longstanding focus on examining the design, implementation, and validation of educational pathways that seek to integrate and harmonize scientific perspectives with those of students. These pathways, often referred to as Teaching/Learning Sequences (TLSs) (Lijnse 1995; Méheut and Psillos

I. Grazia (✉) · O. R. Battaglia · G. Termini · C. Fazio
Department of Physics and Chemistry "Emilio Segrè", University of Palermo, Palermo, Italy
e-mail: ilaria.grazia@unipa.it

O. R. Battaglia
e-mail: onofriorosariobattaglia@unipa.it

G. Termini
e-mail: giulia.termini01@unipa.it

C. Fazio
e-mail: claudio.fazio@unipa.it

I. Testa et al. (eds.), *Connecting Physics Education Research and Practice*,
Challenges in Physics Education, https://doi.org/10.1007/978-3-031-86609-8_5

2004; Psillos Kariotoglou 2015), are highly relevant in physics education research. They serve a dual purpose: as research tools, they enable the study of the effectiveness of proposed activities, while as innovative proposals, they illustrate how learning challenges related to specific topics can be addressed constructively.

When designing a TLS, it is essential to first consider the most suitable approach for representing and describing the phenomena under study to enhance students' understanding, whether they are in school or university. This often requires carefully selecting the appropriate modeling scales, which allow for different perspectives on the phenomena. Only after completing this initial phase, it is possible to proceed with the planning and testing of the TLS. As it is well known, the scales commonly used to analyze a phenomenon are

1. The macroscopic level that encompasses what can be observed directly through the physical properties of matter (such as mass, volume, energy, and force) and its observable manifestations (for instance, a paperclip "floating" on water or a mercury droplet that resists spreading on a surface). This level typically corresponds to orders of magnitude approximately $>10^{-6}$ m (Reif 1965).
2. The microscopic/submicroscopic level that pertains to molecular, atomic, or sub-atomic details (such as crystal structures, molecules, or atoms) that can be detected through specialized instruments and equipment designed to observe interactions at this scale. At this level, phenomena are interpreted based on changes in the composition and structural organization of matter, as well as through molecular interactions. This level generally corresponds to orders of magnitude $\leq 10^{-9}$ m (Reif 1965).
3. The mesoscopic level that serves as an intermediary scale between the microscopic/submicroscopic and macroscopic levels. At this scale, the "particles" under consideration are composed of tens of thousands of molecules, with interactions represented by forces similar in form to those present in the microscopic environment (Battaglia et al. 2021). The size of the mesoscopic particles falls in the range $10^{-9} \div 10^{-6}$ m.

From an educational perspective, it is always useful to determine which modeling scale best supports or enhances the teaching and learning process for the chosen group of students (Besson and Viennot 2004). This is particularly true for a pedagogical approach like the one we want to discuss in this paper that deals with surface phenomena in liquids. The physics of surface phenomena in liquids encompasses a wide range of topics, many of which are often underexplored or overlooked in school and university curricula, and may be unclear and unattractive to the students, also because of the traditional teaching approaches used (Marchand et al. 2011; Berry 1971). Several studies show that from a teaching/learning perspective, it may be relevant to approach these topics based on the explanatory power of the different scales that can be used to model surface tension (Marchand et al. 2011).

A careful consideration of the scales to be used to modeling these phenomena can provide the students distinct insights into the nature and behavior of these complex phenomena. In fact, since surface phenomena in liquids often result from interactions at scales that are not visible to the naked eye, such as microscopic or mesoscopic levels, selecting proper scales can greatly aid students' understanding of how these fundamental interactions influence the observable properties of surfaces.

It may also be effective to design activities that require the use of different scales across various teaching scenarios to better support student learning. To promote conceptual clarity, dealing with macroscopic, mesoscopic, and microscopic scales enables students to explore how surface properties arise from interactions. For instance, studying surface tension through a mesoscopic model bridges atomic-level behavior with macroscopic effects, making intricate scientific ideas more accessible to students, who may initially find an abstract approach, based on the idea of forces acting at a microscopic level, challenging to grasp.

So, the choice of appropriate scales of modeling allows educators to deal with different levels of abstraction. Focusing on a scale suited to students' knowledge allows for simplified models, avoiding excessive detail that may overwhelm learners while still conveying essential mechanisms.

Additionally, the mesoscopic scale offers strong visualization opportunities of the situations in study. Unlike microscopic or macroscopic perspectives, mesoscopic models are well suited to dynamic simulations ran on computer resources easily available in schools and universities, allowing students to see and mentally process these intangible phenomena. By observing, and possibly interacting with real-time simulations, modifying relevant parameters, and seeing the simulation outputs, students can see how particles interact and respond to forces, transforming abstract ideas into visual narratives that enhance their comprehension.

Exposure to multiple modeling scales prepares students for advanced studies and professional work. Learning how to adjust the scale of observation based on complex systems' requirements is a widely applicable skill in fields like physics, engineering, and materials science. By mastering this approach, students gain not only a better understanding of the phenomena studied, but also a versatile analytical skill that they can apply to diverse scientific challenges in the future.

An earlier trial of a pedagogical approach to surface phenomena in liquids was discussed in a previous paper (Fazio et al. 2023). The trial involved high school students in pedagogical activities based on laboratory and modeling. In this paper, we want to discuss how analyzing the results from that prior trial can inform the design of an updated version of the TLS, aimed at undergraduate students. The research here discussed aims particularly to explore how progressively increasing students' active involvement in modeling at different scales and computer-based simulation activities can enhance their scientific learning in meaningful ways.

2 Theoretical Framework and Research Questions

Over the years, extensive research on TLSs has focused on enhancing the effectiveness of science learning. However, improving effectiveness goes beyond developing specific content in a meaningful way. It also involves considering the affective aspects that influence students' well-being and motivation during learning (Rothkrantz 2017;

Etkina et al. 2019). Indeed, the affective component plays a key role in connecting different types of knowledge, simplifying the learning process, and encouraging students to perceive and interpret the surrounding world in new ways.

The approach we aim to apply in designing a student-centered TLS emphasizes exploration, discovery, and information processing led by the students themselves. This approach aligns with the Educational Reconstruction model (Kattmann et al. 1996). This model, rooted in the German pedagogical tradition, is founded on a constructivist epistemology. In constructivism (Bada and Olusegun 2015), learning is seen as an autonomous process in which students reconstruct and adapt their mental models to integrate new experiences. In this view, each student individual conceptions and beliefs are not barriers to learning but are considered valuable starting points for guiding them toward scientific knowledge.

In our TLS design, we applied this model and incorporated active learning methodologies to facilitate an educational reconstruction of the content on surface phenomena in liquids performed also by the students themselves. In such a pedagogical setup, students engage actively with their learning through discovery, information processing, and application. In this context, the quality of the learning environment (Zajda 2021) becomes crucial. Learning environments are regarded not merely as physical spaces but as resource-rich contexts in which each student can navigate freely without rigid instructional constraints. This design approach aims to respect students' unique needs and learning capacities, including their pre-existing conceptions, general thinking patterns, and problem-solving skills. Students are encouraged to collaborate on projects, present and debate their ideas, ask questions, search for answers, and contribute to group discussions, fostering an inclusive and participative learning experience.

The preliminary phase of designing the TLS on surface phenomena in liquids that we aim to discuss here concerns the identification of the general research problem and some research questions to address. Our research problem can be traced back to the wide question "What aspects of the TLS can be considered relevant in promoting students' scientific learning?", which we will be able to answer only after having collected and analyzed the data during the next trial.

To address this question and establish a valid method for evaluating the effectiveness of TLS in promoting scientific learning, it is essential to first ask a more fundamental question: "What does 'promoting learning' really mean in science?".

To answer this last question, a thorough bibliographic review was initially conducted to explore the various characteristics of learning that researchers and educators typically focus on when studying this concept. The study of what is meant in the literature by "promotion of student learning" has revealed many facets and complexities, as well as the lack of a single methodology to describe this process. This led us to build a concept map (see Fig. 1) that highlights three main aspects of learning, which can be called "dimensions" of learning (Marzano 1992), that we consider relevant for a study such as the one we want to conduct. In addition to this, we have identified some sub-dimensions, which describe in finer detail the dimensions of learning and allow us to better analyze how, and in what, a given TLS can be considered effective in promoting scientific learning. More details about the map,

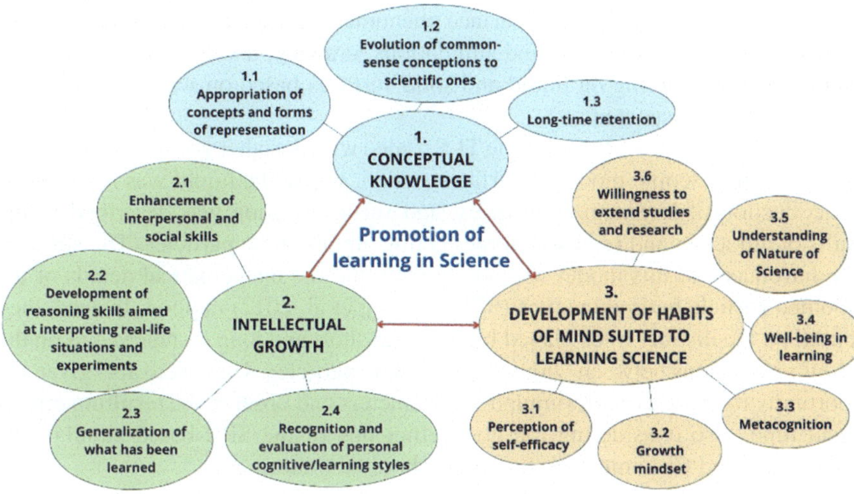

Fig. 1 Concept map highlighting dimensions and sub-dimensions of learning identified through literature review

and the dimensions and sub-dimensions of learning can be found in (Fazio et al. 2023).

As we said before, the main purpose of this paper is to discuss which aspects of the previous TLSs trial provided valuable insights on modeling to inform the design of the updated TLS for university students.

In the following sections, after briefly introducing the main features of the TLSs already trialed, and some of the results obtained from the analysis of the data collected, we will focus on the main changes we have made to the updated TLS to improve its effectiveness.

3 Main Results of the Pilot Trial

In the 2021/22 school year, a 24 h trial on surface phenomena in liquids was conducted with two different groups of students, all attending an upper secondary school with a scientific curriculum. The trial aimed to investigate the differences in learning outcomes of two groups of students experimenting with macroscopic and mesoscopic approach to surface phenomena in liquids, respectively. The experimental nature of both TLSs led to notable improvements in both student groups across several of the sub-dimensions of learning introduced in Fig. 1 (from now on, called "study variables").

As we said, both groups engaged in modeling activities as part of the trial. The group dealing with a macroscopic approach (Group A) were introduced to the general concept of surface tension and to cohesive and adhesive forces on liquid and solid surfaces as the cause of surface phenomena in liquids. They studied books and watched videos of lessons and simulations, drawing, and discussing graphical schemes of forces acting on liquid and solid surfaces based on what is reported in the books/videos inspected.

The modeling approach used in the TLS trialed with the other group (Group B) was based on a mesoscopic model of surface tension, where the liquid was represented as a collection of Lagrangian particles, and surface tension was described as the results of attractive and repulsive forces acting among these particles. The students were introduced to this model without delving into the mathematical details of the smoothed particle hydrodynamics (SPH) (Zhang et al. 2012) method. Instead, they focused on the distinct roles played by forces at short and long distances and on the various interactions between "liquid" and "solid" particles. They were also given the opportunity to adjust certain simulation parameters and observe the resulting effects in real time, also considering the good efficiency of the SPH-based simulations, implemented in PCs commonly found in school laboratories.

The results of the pilot trial, briefly presented in this section, are discussed in more detail in (Fazio et al. 2023). They indicated that the macroscopic approach, which focused on conducting experiments and discussing descriptive and interpretive models based on forces, pressures, and energies, led Group A students to analyze the phenomena primarily by drawing on concepts from their prior chemistry and physics courses, rather than by primarily seeking cause-and-effect mechanisms to explain their observations. Notably, this macroscopic approach provided students with a more robust conceptual foundation compared to the mesoscopic approach when explaining complex situations, such as experiments involving interactions between liquids and solids (e.g., capillary phenomena and meniscus formation). Students following the macroscopic approach were generally able to "explain" their observations more readily, although their explanations were often only descriptions, not rooted on individuating cause–effect mechanisms. Furthermore, referring to the concept map in Fig. 1, students who were involved in the macroscopic approach developed stronger personal and social skills during quantitative experiments (variable 2.1). These skills enabled them to better understand the scientific aspects of interaction and information sharing compared to the group following the mesoscopic approach. Additionally, they showed an improvement in their ability to generalize the skills they acquired (variable 2.3), which is valuable for adopting new methodologies in the school context.

All the activities undertaken by Group B students were supported by mesoscopic-scale simulations that allowed students to "visualize" interactions between the mesoscopic "particles". Rather than relying on prior knowledge and descriptions, students were encouraged to construct explanations and explore cause-and-effect relationships to better understand the phenomena under study. Our analysis indicates that this group of students developed and strengthened reasoning skills for interpreting real-life situations (variable 2.2). Additionally, they showed a greater willingness to pursue further studies and research compared to Group A students, thereby enhancing competencies related to variable 3.6.

We also observed that the mesoscopic approach helped Group B students build confidence in their ability to reason about physical phenomena, even those not directly covered in the TLS. This increased confidence fostered behaviors characteristic of a growth mindset (variable 3.2) (Dweck 2006). This finding suggests that an educational approach centered on constructing and utilizing explanatory models highlighting cause-and-effect relationships can significantly encourage students to reflect on their knowledge and reasoning methods. It shows that a strictly "hands-on" approach may not always be sufficient for mastering complex topics.

Overall, both approaches fostered the construction of knowledge through scientific methods, including collaboration, discussion, and the shared validation of results. Students in both groups reported similar levels of well-being in their learning experience (variable 3.4). These findings are essential, as they show that this type of approach was positively received by both groups and allowed students to re-examine surface phenomena in liquids from multiple perspectives. The students engaged with a complex and varied learning environment that actively supported their development of deep understanding and self-awareness, while also promoting a more effective attitude toward problem-solving.

However, the analysis of the data collected during the TLSs trialing revealed some limitations, which will be examined in the following section. Based on these findings we designed an updated TLS, which will be tested with undergraduate students. This will incorporate the strengths of the two approaches already used, extend the time spent on modeling activities carried out through simulations, and engage students in even simple tasks to understand the algorithms on which such simulations are based.

When we planned to design the updated TLS, one key aspect on which we focused was the decision to deepen the description of surface phenomena in liquids based on the mesoscopic model.

As we mentioned before, introducing mesoscopic models in the explanation of surface phenomena in liquids holds significant educational and scientific value. These models bridge the gap between microscopic and macroscopic perspectives, allowing students and researchers to understand complex processes at a scale that captures essential details without the overwhelming specificity of atomic/molecular models or the oversimplifications of purely macroscopic approaches. Surface phenomena in liquids involve interactions that occur at multiple scales, from atomic to macroscopic levels. Mesoscopic models help learners appreciate the intermediate level, where collective particle behaviors and interactions can be observed without going into the extreme detail of individual atomic/molecular interactions. This middle ground is particularly useful for phenomena at surfaces, where properties like energy, surface tension, and molecular arrangement reveal essential features that are often hidden in purely macroscopic descriptions. The mesoscopic scale enables effective visualization of processes that are difficult to illustrate at atomic/molecular or bulk levels.

Through computer-based simulations, based on simplified models for educational purposes, students can observe the patterns and fluctuations that emerge at surfaces, such as the behavior of particles during the formation of a liquid drop or the rise of a liquid in a capillary tube. This level of visualization enhances conceptual clarity, helping students form intuitive understandings of processes like liquid–gas interface behaviors, and thin film formation, which are otherwise challenging to comprehend.

Using mesoscopic models for algorithm development offers a balanced approach to understanding complex systems without the computational demands of microscopic models. One of the main advantages is reduced computational complexity: mesoscopic models simplify particle interactions by focusing on group behaviors rather than individual atoms or molecules, allowing faster, more scalable simulations. This makes it feasible to model larger systems and longer timescales without sacrificing key details. Furthermore, mesoscopic models facilitate visualization, making complex processes easier to interpret and communicate, especially in educational contexts.

Working with mesoscopic models encourages students to think analytically and critically about complex systems. Since these models often involve simplifications and approximations, students learn to evaluate the assumptions involved and consider the model's limitations. This practice develops important scientific skills, such as model evaluation, hypothesis testing, and an appreciation of the complexity inherent in real-world phenomena. As students engage with mesoscopic models, they develop a more nuanced understanding of surface phenomena in liquids and learn how to approach complex problems systematically.

We think that mesoscopic models could offer students a new perspective on surface phenomena in liquids. By facilitating visualization, simplifying complex concepts, bridging theory with practical applications, these models could contribute significantly to physics education and research. The mesoscopic approach, introduced through the computer-based simulations, therefore, could be an instructional tool useful in fostering a comprehensive, nuanced understanding of surface phenomena in liquids.

4 From the Trial TLSs to the Updated TLS

In the light of the conclusions of the first trial of the TLSs, which highlighted the need to enhance the use of interactive simulations in education, linking the results achieved through computer simulations to the experimental ones, we now provide an overview of how to develop and validate a TLS with university students on surface phenomena in liquids strongly based on active learning methodologies.

The new trial is conducted on students attending the second and third years of three different engineering courses of the University of Palermo, who have been offered to attend the trialing as an integrative teaching activity authorized by their respective degree courses (2–3 recognized university credits). Students are involved in an extracurricular activity, lasting 30/35 h. The research focuses on how to enhance the processes of active knowledge construction, favoring the appropriation of physical concepts by students, rather than attending a simple transmission of knowledge. In the literature, several processes have been suggested to describe the conceptual changes that happen in the students when they are involved in physics lessons and tend to approach a branch of knowledge that they had never faced until then. The peculiarity of the updated TLS lies not only in proposing physical content not expected within the curricula, but in proposing them to students in an affectively and cognitively engaging way.

In this section, we also examine the main limitations we encountered while experimenting with the first version of the TLSs. We present and discuss selected excerpts from the collected data, highlighting the key adjustments we believe can further strengthen the updated TLS.

Subsequently, we introduce the strategic measures and refinements implemented in designing the updated TLS activities, with the goal of enhancing the learning experience and addressing identified areas for improvement. These modifications are intended for incorporation in the upcoming phase of experimentation, aiming to create a more effective and engaging educational environment.

One of the key issues identified was the insufficient time allocated to simulation activities involving a mesoscopic approach. According to the students, this limited time hindered their ability to fully grasp the topics addressed and the material and methods used. Below we report some student feedback comments on the computer-based simulation experience, highlighting the challenges related to the timing of these activities.

Student 1 feedback on the time dedicated to computer-based simulation activities:

> I often felt rushed when using the computer simulations. With such a short amount of time, I could not explore or experiment with the tools as much as I needed to truly grasp the complex concepts.

Student 2 feedback on the time dedicated to computer-based simulation activities:

> Due to the restricted timeframe, I was usually only able to complete the basic steps in each simulation, which made it difficult to test different scenarios or fully benefit from the interactive experience.

Student 3 feedback on the time dedicated to computer-based simulation activities:

> With so little time dedicated to the simulations, I struggled to develop the necessary technical skills. This left me unable to focus properly on learning the scientific concepts the simulations were intended to teach.

These comments suggest that students encountered challenges with the limited time provided for running the computer-based simulations and make sense of the results. The brief timeframe made it difficult for them to fully explore the tools and experiment with different scenarios, which seemed to impact their grasp of complex concepts. Furthermore, the restricted practice time posed difficulties in developing essential technical skills, which may have diverted their focus from the primary scientific objectives. This feedback indicates that allowing more time for simulation activities, and for the discussion and contrast of the results, could better support students in building both their technical abilities and conceptual understanding.

Another limitation related to the choice of student participants; we found that the age and foundational knowledge of the fourth-year scientific high school students involved in the pilot trial may have restricted their ability to gain a productive understanding of the algorithms being studied. The following student feedback highlights the issue just discussed.

Student 1 feedback on the previous knowledge required to manage surface phenomena in liquids through computer-based simulation effectively:

> I found it challenging to fully grasp the concepts related to surface phenomena in liquids. My educational background in such scientific topics and tools is limited, which made it difficult for me to understand both the phenomena and the computer-based simulations.

Student 2 feedback on the previous knowledge required to manage surface phenomena in liquids through computer-based simulation effectively:

> I encountered some difficulty in achieving a deep understanding of surface phenomena in liquids. This limited knowledge base presented challenges in utilizing the computer-based simulations effectively, which in turn affected my overall comprehension of the methodologies involved.

Student 3 feedback on the previous knowledge required to manage surface phenomena in liquids through computer-based simulation effectively:

> Due to my scientific background, I sometimes felt I lacked the skills needed to accurately interpret surface phenomena in liquids. Consequently, I found it difficult to engage with the methodologies, including computer-based simulations, which necessitated a more advanced understanding of both the tools and the underlying scientific principles.

The previous excerpts illustrate the difficulties students faced in comprehending surface phenomena in liquids and/or the functioning mechanism of computer-based simulations, largely attributed to their educational background. They convey a sense of struggle in understanding complex topics and new learning tools and strategies, suggesting that their foundational knowledge may not have adequately equipped them for facing these challenges. This feedback highlights the necessity of adapting educational strategies to address the varied needs of learners.

While school students may benefit from similar strategies, we think that the educational background of undergraduate students can make them more receptive to and capable of engaging with these types of learning activities. University students typically possess a deeper understanding of foundational concepts due to their previous academic experiences, allowing them to approach complex topics with greater confidence even when they are asked to deal with new tools and methodologies. This prior knowledge enables them to make connections between new information and existing frameworks, facilitating a more meaningful learning experience. Moreover, university education often emphasizes independent learning and critical thinking skills, which are essential when tackling advanced subjects such as surface phenomena in liquids and computer-based simulations. University students are encouraged to explore topics in greater depth, fostering an environment where they can experiment, ask questions, and challenge assumptions. This level of engagement not only enhances their comprehension but also encourages the development of problem-solving skills that are crucial in real-world applications. In contrast, younger students may still be developing the skills needed for independent learning and may require more structured guidance to navigate complex concepts. While they can certainly benefit from similar teaching strategies, the implementation may need to be adapted to suit their developmental stage, with a greater emphasis on scaffolding and guided support.

In response to these findings and considerations, we decided to develop a new version of TLS that would integrate the strengths of the two teaching approaches previously experimented, and to experiment it with university students. This revised TLS will allocate more time to simulation activities, allowing for a deeper exploration of concepts.

To improve the TLS based on the identified limitations, several specific questions and considerations emerged that guided the redesign process.

The TLS taken by Group B students dedicated limited time to mesoscopic simulations, which left students struggling to grasp the concepts thoroughly. This shortcoming highlighted the need to reassess time allocation for simulations. In the updated TLS, we significantly extend simulation time, giving students a more comprehensive experience with mesoscopic models and allowing them to interact longer with these tools. By doing so, students can explore more in-depth various simulation parameters, observe outcomes more closely, and refine their understanding through experimentation. This increased exposure aims to foster a deeper understanding of complex content, bridging the gap between theory and application.

Moreover, we discussed the role of preliminary knowledge in engaging students with the simulations effectively. A clear understanding of the prerequisite knowledge emerged as a vital consideration. Students need a foundation in basic scientific concepts to make sense of the algorithms and simulation tools involved in the TLS. Therefore, the updated TLS will include a preparatory module or review segment where essential concepts, such as basic algorithmic logic, mathematical modeling, and data interpretation, are covered briefly. This segment will allow students to build the needed context before diving into more complex topics, setting them up for a more productive learning experience.

Another relevant question emerged from our consideration about the role of computer-based simulation in education. For many students, understanding algorithms involves engaging with overly complex tasks that can be difficult to decode. By breaking down these algorithms into simpler, step-by-step activities, the updated TLS aims to demystify them. A scaffolding approach intends to build confidence and understanding, allowing students to transition smoothly from simple concepts to more intricate processes.

In Appendix A, we include an example of a worksheet that will be provided to students as part of the updated TLS.

5 Conclusion

As discussed in the introductory section, considering the explanatory power of different modeling scales can be highly relevant for teaching and learning complex scientific topics as surface phenomena in liquids. We hypothesized that choosing an appropriate modeling scale to introduce this topic could significantly enhance the teaching and learning processes at both the school level—as observed in our pilot trial—and potentially at the university level. Based on this hypothesis, we designed a TLS that integrates the strengths of both macroscopic and mesoscopic approaches. This design aims to address not only the understanding of surface phenomena but also the broader question of what it truly means to learn.

The pilot TLSs highlighted strengths in both theoretical instruction and computer-based simulations. In the updated TLS, we aim to strike a balance between these approaches by alternating between theory and practice in a way that reinforces learning. For example, after a theoretical lesson on a specific concept, students might immediately engage in a practical activity that applies that concept in the simulation. This iterative cycle of theory and practice ensures that students understand the relevance and application of the material in real-time, strengthening retention and mastery.

By addressing these questions, the updated TLS aims to foster a more engaging and accessible learning experience that meets the diverse needs of students. This process involves carefully balancing theory with laboratory activities, extending the time allocated for simulations, and scaffolding tasks to gradually introduce complex concepts. Through this approach, we believe students will be better equipped to understand and apply the algorithms, making the learning experience both more enjoyable and effective.

Appendix A

Simulated Experiment: Formation of a Drop of Water in the Absence of Gravity

Note: The two screenshots below were generated in MATLAB using output obtained by running a script based on the SPH (Smoothed Particle Hydrodynamics) algorithm. In these images, pressure forces are shown in yellow, while liquid–liquid interaction forces are represented in red. These combined simulate molecular interactions in liquids, acting as attractive forces over longer distances and as repulsive forces at shorter distances.

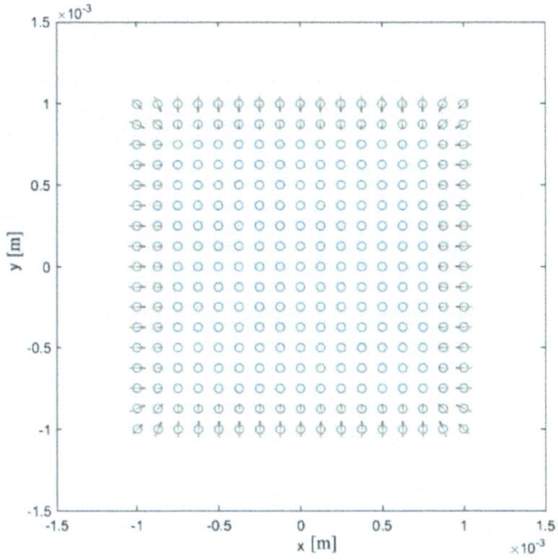

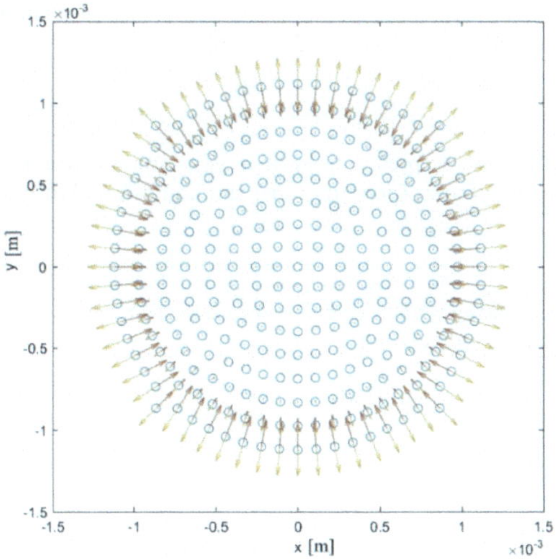

Analyze the behavior of the simulated liquid for a liquid–liquid interaction value of 3.0.

Individual Student Observation

1. After becoming familiar with the simulated system, try pausing the simulation shortly after it starts and, using the MATLAB function plot (…), create a graph of it. The resulting image will depict a square distribution of particles. How do you think pressure force and interaction forces (or interparticle forces) affect the system?

 ____

 ____

2. Which aspects of the mesoscopic model did you find most useful for understanding the forces acting on the drop?

 ____

 ____

Small Group Discussion

Now discuss with your group the answers you gave to points 1. and 2., and compare them with the answers given by your group-mates.

3. After discussing with your group, describe if your understanding of the phenomenon has changed or remained the same.

 ____

 ____

4. Did any different interpretations arise in your group discussion regarding the effect of the two forces (pressure force and liquid-liquid interaction force)?

 ____

 ____

5. How did discussing with your classmates help you improve your understanding of the simulated system?

 ____

 ____

Large Group Discussion

Now, choose a representative within your small group to present your conclusions to the large group. After listening to the representatives of all the small groups, start a large group discussion to compare your different positions.

6. Are the conclusions reached by the large group discussions compatible with your predictions? And with those of the small group?

 ____

 ____

7. At the end of this activity, do you really agree with what was agreed by the large group? Explain your position as best you can.

 ____

 ____

References

Bada SO, Olusegun S (2015) Constructivism learning theory: a paradigm for teaching and learning. J Res Method Educ 5(6):66–70
Battaglia OR, Gallitto AA, Termini G, Fazio C (2021) A study of capillarity phenomena by using a computer-based simulation approach. Eur J Phys 42(5):055704
Berry MV (1971) The molecular mechanism of surface tension. Phys Educ 6(2):79
Besson U, Viennot L (2004) Using models at the mesoscopic scale in teaching physics: two experimental interventions in solid friction and fluid statics. Int J Sci Educ 26(9):1083–1110
Dweck CS (2006) Mindset: the new psychology of success. Random house, New York
Etkina E, Brookes DT, Planinsic G (2019) Investigative science learning environment: when learning physics mirrors doing physics. Morgan & Claypool Publishers, San Rafael, CA
Fazio C, Gallitto AA, Galiano CG, Giarratano G, Grazia I, Termini G, Battaglia OR (2023) An approach to research-based design of teaching-learning sequences in the context of physics education: theoretical frameworks, pedagogical methods, and examples of Data Analysis. Il Nuovo Cimento 46:199–227
Kattmann U, Duit R, Gropengießer H, Komorek M (1996) Educational reconstruction–bringing together issues of scientific clarification and students' conceptions. In: Annual meeting of the national association of research in science teaching (NARST). St. Louis, pp 1–20
Lijnse PL (1995) "Developmental research" as a way to an empirically based "didactical structure" of science. Sci Educ 79(2):189–199
Marchand A, Weijs JH, Snoeijer JH, Andreotti B (2011) Why is surface tension a force parallel to the interface? Am J Phys 79(10):999–1008
Marzano RJ (1992) A different kind of classroom: teaching with dimensions of learning. Association for Supervision and Curriculum Development, 1250 North Pitt Street, Alexandria, VA 22314
Méheut M, Psillos D (2004) Teaching–learning sequences: aims and tools for science education research. Int J Sci Educ 26(5):515–535
Psillos D, Kariotoglou P (eds) (2015) Iterative design of teaching-learning sequences: introducing the science of materials in European schools. Springer
Reif F (1965) Fundamentals of statistical and thermal physics. Waveland Press
Rothkrantz L (2017) Affective didactic models in higher education. Int J Hum Capital Inf Technol Prof (IJHCITP) 8(4):50–66
Zajda J (2021) Globalisation and education reforms: creating effective learning environments, vol 25. Springer Nature
Zhang M, Zhang S, Zhang H, Zheng L (2012) Simulation of surface-tension-driven interfacial flow with smoothed particle hydrodynamics method. Comput Fluids 59:61–71

References

Effectively Teaching the Physical Foundations of the Greenhouse Effect: Evolution of a TLS

Alessandro Salmoiraghi, Stefano Toffaletti, Marco Di Mauro, Camilla Fiorello, Tommaso Rosi, Eugenio Tufino, Pasquale Onorato, and Stefano Oss

Abstract A scientific understanding of the greenhouse effect (henceforth, GHE) is crucial to assess the human role in this phenomenon. This paper presents the evolution of a teaching–learning sequence (TLS) focused on the physical basis of the GHE, which consists of a set of experiments and laboratory activities. The sequence was designed, implemented, and validated using an iterative approach in accordance with the Model of Educational Reconstruction (MER). The latest redesign of the sequence was influenced not only by the results of the previous ones, but also by the evolution of the educational context and researchers' awareness. In particular, the redesign aimed to integrate the science discipline with the more typical aspects of science literacy. The goal is to improve citizens' scientific awareness; an aspect that can influence their decisions and sensibility to face global challenges, such as climate change.

1 Introduction

In recent years, issues related to the GHE and climate change (henceforth, CC) have become increasingly popular (Stocker 2013). In this area, physics education has a primary role in the training of students, not only strictly regarding content (the physical bases), but more importantly scientific literacy in a broader sense (training students as future citizens and decision-makers) (Pachauri and Meyer 2014).

Our research group has been working on these issues for years, developing TLSs on the physical basis of GHE (Besson et al. 2010a; Onorato et al. 2011, 2021a; Toffaletti et al. 2022, 2023). Over time, the motivations that guided this research have changed. As a result, teaching choices have been influenced time after time by several factors, both internal and external, that have driven the redesign of the TLS.

A. Salmoiraghi (✉) · S. Toffaletti · M. Di Mauro · C. Fiorello · T. Rosi · E. Tufino · P. Onorato · S. Oss
Department of Physics, University of Trento, Trento, Italy
e-mail: a.salmoiraghi@unitn.it

I. Testa et al. (eds.), *Connecting Physics Education Research and Practice*, Challenges in Physics Education, https://doi.org/10.1007/978-3-031-86609-8_6

In this paper, we will briefly retrace this evolution, shortly reviewing its stages and describing the motivations that guided our design choices. The main aim of this work is to describe the different versions of a TLS on global warming, by discussing how it was modified during the years, and to analyze the results we obtained so far to answer the following research questions:

RQ1: How did the iterative process of validating and testing the TLS change its contents and learning objectives? How have the proposed activities changed?
RQ2: Starting from the changed perception of climate change by society and students, how do these and other factors influence the process of redesigning the teaching sequence?
RQ3: Does TLS manage to be effective in terms of science learning? Does the activity succeed in changing students' engagement with the problem of climate change?

2 Methodological Frameworks

The methodological framework used to design the first sequences was the three-dimensional approach of Besson et al. (2010b), which was later integrated with Duit's Model of Educational Reconstruction (MER) (Duit et al. 2012). The latter requires research to move contextually on three domains, which mutually influence each other in the development of the educational project. They are:

- Clarification and analysis of science content: In this phase, we analyzed the science content presented in high school textbooks (Toffaletti et al. 2022), the identification of the key ideas of GHE physics (Onorato et al. 2021a), and the construction of the content structure for teaching.
- Research on teaching and learning: In this area, we provided for the analysis of the results of pre- and post-tests administered to students who, in the various years, attended the university courses in which TLS was being tested (Besson et al. 2010a; Onorato et al. 2021a; Toffaletti et al. 2022, 2023); we identified their preconceptions, misconceptions, interests, and attitudes, creating (first through a principal components analysis (Fiorello et al. 2023; Liu 2021) and later through a cluster analysis) a map of their mental models of the phenomenon.
- Design and evaluation of teaching and learning environments: Analyses of the learning outcomes of the various versions of TLS were the guidelines for finding the conceptual steps whose understanding needed to be strengthened (Onorato et al. 2021a; Toffaletti et al. 2022, 2023).

Over the years our research group has investigated the three components of the model, designing updated versions of the sequence time after time, in accordance with an iterative process typical of this framework, because "somewhat recursive procedure is needed to re-construct an appropriate content structure for instruction" (Duit et al. 2012).

The Design and Redesign Model to enhance understanding and learning (DREM) represents a synthesis of the MER (Duit et al. 2012) and the three-dimensional approach proposed in Besson et al. (2010b). The model originates from the idea of design–redesign proposed in many papers, such as those stemming directly from Besson's work (cf., e.g., (Onorato et al. 2017), where the cycle of design, implementation, evaluation, and redesign is reported: the initial TLS, rooted on a careful textbook analysis and on research results on students' difficulties, was modified according to the results of previous testing).

The iterative redesign process of Teaching–Learning Sequences (TLS) plays a critical role in refining educational interventions by integrating theoretical insights and practical applications, as deeply explored by Psillos and Kariotoglou (2016). This process is particularly significant in the context of teaching scientific concepts through inquiry-based approaches. The iterative design process involves continuous cycles of design, implementation, evaluation, and refinement, ensuring that the TLS evolves to meet the learning needs of students more effectively. This approach emphasizes both the integration of formative assessment to support iterative refinement (Spyrtou and Kariotoglou 2016) and the continuous improvement based on empirical data from classroom practices, aligning the TLS with educational goals and students' learning needs.

The iterative process generally involves multiple iterations, each informed by students' responses and the need to address specific misconceptions. Some exempla of TLS aimed at introducing inquiry-based learning which integrates hands-on experiments and scaffolded inquiry activities to enhance understanding of key concepts were reported in (Psillos and Kariotoglou 2016). These examples underscore the importance of the iterative redesign process in developing effective TLS. By continuously refining the sequences based on classroom feedback and educational research, educators can create learning experiences that are both scientifically rigorous and pedagogically sound. This approach not only enhances students' understanding of complex scientific concepts but also fosters a deeper engagement with the material.

In summary, the iterative process allows educators to fine-tune the sequence, ensuring that it not only conveys the scientific concepts effectively but also engages students in meaningful inquiry-based learning.

The model shown in Fig. 1, starting from the idea of iterative redesign process, highlights the typical elements of the MER (content analysis to identify fundamental concepts and structures, clarification of subject matter and analysis of educational significance, critical analysis of the scientific content in view of its reconstruction for teaching, elementarization) and those of Besson's three-dimensional approach (analysis of didactic research on student conceptions and previous TLSs on the topic, overview of the usual treatments (textbooks, common teaching practices), critical analysis of the scientific subject, including its historical development, together with its practical applications).

The sequences, that have been designed and re-designed over the years following this philosophy, will be explained in the following paragraphs.

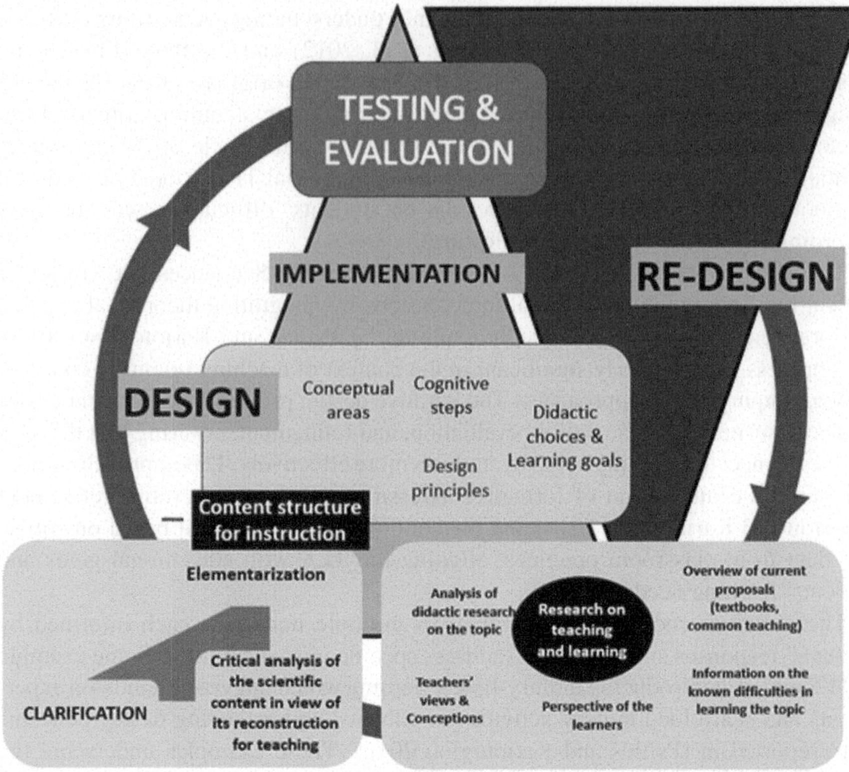

Fig. 1 A scheme of the design and redesign model to enhance understanding and learning (DREM), which synthesizes elements from both the MER model and three-dimensional approach of Besson et al. The model includes content analysis, educational significance, critical analysis, and elementarization from the MER model, along with didactic research, usual treatments, and historical and practical analysis from approach of Besson et al.

3 Previous TLSs

Over the past two decades, numerous studies have investigated students' understanding of GHE and global warming across various educational levels, uncovering a persistently limited knowledge and inadequate explanation of these processes, with difficulties spanning all ages and educational stages. Version 0 (Besson et al. 2010a) of the experimentation had highlighted that students do not adequately assess the phenomenon of infrared radiation (IR) emission as an energy emission mechanism, tend to assign absolute values to optical properties without considering their wavelength dependence, confuse transient phases with stationary state conditions, and show difficulties in considering the interdependence of all factors and phenomena involved in Earth's energy balance.

From the analysis of well-known concept inventories (Keller 2006) administered to many students, further insights into their preconceived ideas have emerged. For instance, some believe that the GHE is due to an increase in incoming solar radiation and a decrease in outgoing flux, confuse concepts associated with the GHE, global warming, and the ozone hole are convinced that solar radiation peaks in the ultraviolet spectrum and use mental models that include permanent radiation entrapment, entrapment through reflection, and entrapment of gases and pollution. Moreover, many think that the Earth's surface primarily emits energy during nighttime hours.

A more general study on CC (Jarrett 2013) has highlighted difficulties in identifying greenhouse gases, an inadequate understanding of Earth's energy balance and blackbody radiation, and a lack of comprehension of the nature of the interaction between electromagnetic radiation and atmospheric gases. Finally, research conducted by the physics education research group of the University of Bologna (Tasquier 2013) has shown that while students do not find the concepts of absorbance, transmittance, and reflectance problematic, they struggle to incorporate the concept of emission into their reasoning and to distinguish between the concepts of radiation and heat, failing to think in terms of energy balance.

4 First Redesign

4.1 Sequence Design

A critical analysis of the results of previous research has identified five cognitive steps essential for a better understanding the GHE. Every step of the TLS is linked to specific experimental activities (Moggio 2017, Table 1). Test with university students has shown that, after following the teaching–learning sequence, students have improved their understanding of the GHE, recognizing the emission of IR radiation by bodies and the dependence of transmittance on the frequency of electromagnetic radiation. They also identified transient and stationary state temperature phases and considered different processes in the behavior of a greenhouse, avoiding the idea of "trapped solar radiation" and recognizing the active role of Earth in emitting IR radiation.

The results from the trials of previous TLSs with secondary school students (ages 15–18) showed a significant reduction in explanations based on thermal insulation and an increase in the use of the distinction between absorption and emission for visible and infrared radiation in explanations of the GHE. However, only about 50% of the students correctly used these concepts after the didactic sequence.

The redesign of the sequence included the definition of detailed conceptual areas, to refine the cognitive steps and related activities (see Table 2). The didactic choices in this phase included the design of experimental activities on each conceptual sub-area, quantitative experiments for accurate data analysis, the use of smartphones

Table 1 Cognitive steps and corresponding activities related to understanding temperature, radiation, and their effects—Version 0. Version 0 of the TLS lasted 12 h and included a module on atmospheric physics, which was eliminated in subsequent versions

Cognitive steps	Activities
Recognize and explain the stationary condition of the temperature of objects exposed to solar radiation or a lamp (energy balance and radiative equilibrium)	Measurements of the temperature over time of metal cylinders of different colors exposed to solar radiation or under a lamp
Discriminate between concepts of heat and radiation and recognize that objects emit thermal radiation	Experiments on the emission of infrared radiation from bodies at different temperatures with a radiometer
Distinguish visible and infrared radiation and the behavior of a material based on different types of incident radiation	Observation of the infrared radiation emitted by a remote control with a digital camera
Combine the previous elements to understand the radiative nature of the greenhouse effect in a model	Qualitative analysis of the spectrum of radiation
Understand the GHE on Earth and global warming from an energy flow model developed for modeling the GHE	Experiments with a model where the behavior of a small greenhouse placed under solar radiation was studied during an outdoor activity

as measuring tools and interfaces for tests (BYOD), the integration of real experiments with interactive simulations, the discussion of the microscopic aspects of the observed macroscopic phenomena, and the organization of student work in small groups. The Predict-Observe-Explain (POE) strategy (White and Gunstone 1992) was implemented in every conceptual step, requiring students to make predictions, conduct experiments, and interpret results, comparing predictions with observations and experimental data.

On the practical side, some experiments already included in the previous version of the sequence have been revised. Specifically, the PASCO instrumentation (https://www.pasco.com/products/lab-apparatus/thermodynamics/radiation/td-8855) used in the experiments with the radiometer and the Stefan–Boltzmann lamp has been replaced with more readily available materials and the smartphone as measurement instrument, with the goal of more widespread implementation of the BYOD strategy.

4.2 Sequence Summary

The sequence, lasting in total 12 h, begins with the exploration of light phenomena through an experiment with dishwashing liquid. A collimated beam of light from a monochromatic LED source penetrates a container filled with green liquid soap.

Table 2 Conceptual areas and corresponding activities related to understanding temperature, radiation, and their effects—First redesign

Conceptual areas	Subareas	Activities
(1) Heat and temperature	a. Heat and temperature b. Heat propagation c. Microscopic interpretation	Classical experiments of thermology, activities on the microscopic model, interactive construction of heat and temperature concepts
(2) Electromagnetic spectrum and black body	a. Energy of the spectrum bands b. Stefan–Boltzmann law c. Wien's law	Observation of light sources on a diffraction grating, measurement of Stefan–Boltzmann law with an incandescent lamp, observation and measurement of the spectrum at increasing temperatures, PhET simulations on blackbody radiation
(3) Radiation–matter interaction	a. Reflection b. Refraction c. Absorption d. Elastic and inelastic scattering e. Selective transparency f. Effects of concentration g. Selective transparency and IR radiation h. Microscopic model	Experiments of geometrical optics and Snell's law, experiments on absorption, PhET simulation of Beer's law (wavelength and concentration), observations with a thermal camera, demonstration on elastic and inelastic scattering, PhET simulation on radiation–matter interaction
(4) Objects exposed to radiation, stationary state, and energy balance	a. Radiative balance b. Stationary states c. Gray bodies: emissivity and absorbance	Measurement of solar constant, stationary states with the "hydrodynamic model," Leslie's cube, absorbance of objects of different colors
(5) Energy balance of Earth and GHE	a. Greenhouse gas properties b. Albedo c. Feedback d. Construction of models to estimate average global temperature	
(6) Global warming analysis	a. Feedback variation effects b. Climate, pollution, and extreme weather events	Larsen effect with two smartphones and demonstration of modified stationary states with the "hydrodynamic model"

Students are tasked with observing reflection, refraction, absorption, and diffusion—fundamental concepts that pave the way for more in-depth investigations.

The introductory experience with the liquid soap (Onorato et al. 2018) allows students to approach the subject through observations and group discussions. From this preliminary observation, the investigation into absorption begins, proposed through the central experiment of the sequence: the experiment on Beer's law. The structure of the apparatus is similar to the previous experiment, but the container is replaced by a beaker into which the soap is progressively introduced, and, using the light sensor of a smartphone, the intensity of the radiation transmitted through the soap is measured at each addition. The quantitative analysis of the data completes the experience, from which Beer's law is derived. The comparison of data from different groups, equipped with LEDs of different colors, allows for the observation of the selectivity of absorption, a key concept in the GHE.

Changing perspective, the observation through a thermal camera of infrared radiation emitted by a hot body, interposing slabs of different materials such as steel, silicon, and plexiglass, leads students to discuss selective transparency. The use of a PhET simulation (Beer's law) (https://phet.colorado.edu) generalizes the law found and brings to the deduction of the dependence of the transmitted light intensity on the medium concentration. The lesson concludes with a discussion on the energy balance that highlights an apparent violation of the principle of energy conservation.

The second investigation, focusing on energy, begins with the observation of an energy deficit in the Beer's law experiment, between the energy entering the system (incoming radiation) and that measured as outgoing (transmitted radiation). To address this issue, the discussion moves to a microscopic level with the PhET simulation (molecules and light) (https://phet.colorado.edu), shifting the focus to the investigation of the diffusion phenomenon.

An experiment on elastic diffusion with milk dissolved in water clarifies the energy balance: the missing outgoing energy is attributed to light scattered by the medium, which escape the sensor (that only collects transmitted radiation). However, a further experiment on inelastic diffusion with fluorescein reintroduces the energy problem, as the scattered photons have lower energies than the incident ones. After a collective discussion, students hypothesize that the missing energy can be measured as an increase in the medium's temperature.

To validate this hypothesis, the experiment with two aluminum plates, one blackened and one polished, is conducted: at regular intervals, students measure the temperature of the plates exposed to an incandescent light bulb. This experiment not only validates the thesis but also introduces concepts such as albedo, steady-state temperature, and radiative equilibrium. To explain these concepts, a model is constructed, the embryo of what will be developed in the second part of the sequence. The model is illustrated through the analogy of the leaking bucket (hydrodynamic model) (Moggio 2017), which concretizes the stationarity of the radiative equilibrium.

In the conclusion of the experimental part, students explored the emission of thermal radiation through the Stefan–Boltzmann lamp experiment, which led to the formulation of the law of the same name. The experiment was extended with the

Leslie cube, introducing the concept of emissivity, and with the use of the spectroscope, combined with the PhET simulation on blackbody radiation (https://phet.colorado.edu), to correlate the type of radiation emitted to the temperature of a body.

Subsequently, the class constructed a climate model, starting with the repetition of the plate experiment to calculate the solar constant and applying the concept of albedo. They created an initial version of the climate model parallel to that of the plate, replacing the lamp with the Sun and the plate with the Earth. However, this model proved inadequate, providing a global average temperature not in line with existing data.

After discussing Beer's law experiment, the class hypothesized that the atmosphere has to play a major role in the model. Students then independently developed a Sun–Atmosphere–Earth model, which includes the GHE, allowing for the calculation of a global average temperature consistent with the data.

The sequence concluded with reflections on positive and negative feedback systems, the concept of climatic chaos, and a PhET simulation on the GHE (https://phet.colorado.edu), providing a holistic understanding of climatic processes and their modeling.

4.3 Sequence Assessment

During the trial of the sequence, which involved 86 graduated students, we administered multiple-choice and open-ended questionnaire before each experimental activity and at the end of the whole sequence. From the analysis of the test results (Fig. 2), the sequence has been effective in addressing the conceptual areas and their corresponding sub-areas identified during the design phase and reported in Table 2. However, from a qualitative analysis of the open-ended question two critical sub-areas were identified. Specifically, students exhibited persistent difficulty in defining the concept of selective transparency in the context of the atmosphere and in outlining the phenomenon of thermal emission.

5 Second Redesign

5.1 Motivation

Between 2021 and the present, our research group has been working on the second redesign of the TLS. It was driven by some "internal" motivations classically present in an iterative process:

- the outcomes of the analysis of previous experiments (Onorato et al. 2021a; Toffaletti et al. 2022);

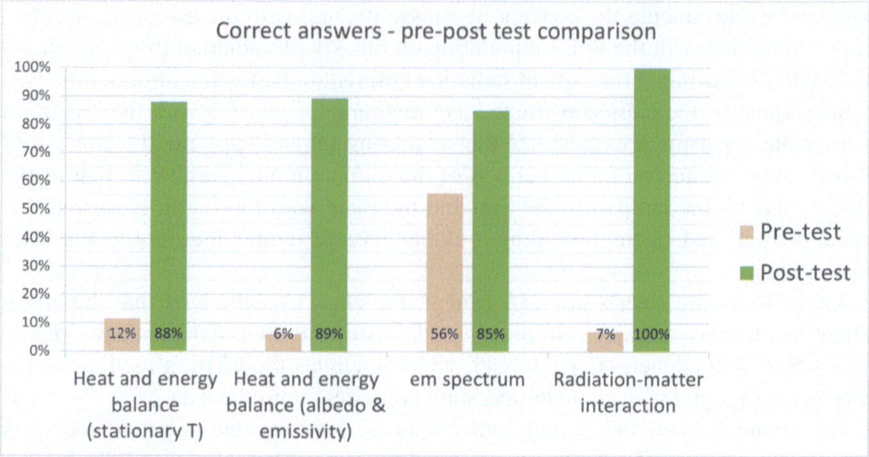

Fig. 2 Comparison between pre-test and post-test results in selected conceptual areas from a class of 86 graduated students. The pre-tests were administered before each experimental activity, the post-test at the end of the sequence

- the analysis of textbooks (Toffaletti et al. 2022); and
- the investigation of students' mental models (Fiorello et al. 2023).

Alongside these, however, we must identify some others that are "external": new compared to the past and therefore crucial to the design adjustments that have been implemented. These are summarized as follows:

- In 2020, the teaching of "educazione civica" (civic education) was reintroduced in Italian schools: one of the conceptual cores of this new teaching is in fact that of "sustainable development," which echoes the thirteenth goal of the UN Agenda 2030 regarding climate action.
- The increasing diffusion, among the public and especially among the younger generation, of the reality and the urgency of the climate issue (Pachauri and Meyer 2014).
- The consequent increased sensitivity on this topic on the part not only of students, but also of teachers themselves.
- According to some recent studies (https://counterhate.com/research/new-climate-denial), CC denial is evolving alongside the preconceptions held by students. In the past, traditional climate deniers often denied the rise in global temperatures and its anthropogenic origins, or they cited phenomena like the ozone hole, which had long been the focus of media attention. However, the new wave of climate deniers has shifted their stance to a "New Denial" of climate impacts. This includes arguments such as "Climate solutions won't work," "Climate science is unreliable," "The impacts of global warming are beneficial or harmless," and "The climate movement is untrustworthy."

All these aspects helped us to identify what were the key aspects whose understanding should be fostered, and how to adjust the sequence of introducing the topics to make the second redesign of the sequence.

The second TLS redesign work thus involved three main interventions: elementarization, construction of content to teach, and update of some experiments.

5.2 Elementarization

First, we used all the phases of previous research (scientific context analysis, textbook analysis, students' misconceptions understandings, and their mental models) to identify the key ideas related to the physical basis of GHE and their level of understanding related to previous experiments.

5.3 Content Structure

Hence, this provided us with the opportunity to construct the new content structure to teach. We focused on explicitly clarifying all the conceptual steps required to understand the phenomenon and their function in the overall structure, without increasing the duration of the sequence, which remained 12 h.

In the first phase, a naive model of the radiative budget between the Earth and the Sun is created, in which the atmosphere is not included. The construction of this "first model" requires an understanding of the concepts related to the interaction of electromagnetic (EM) radiation with matter from the macroscopic point of view (reflection, refraction, absorption, and scattering) and from the microscopic one (Wien's law, Stefan–Boltzmann's law, and radiative budget). The model thus created provides a result (in terms of equilibrium temperature) that is far from the actual experimental mean value. This allows us to point out that it is essential to include the atmosphere in the model.

The second step involves the creation of a "second model" of radiative balance, which also includes the atmosphere and the concentration of greenhouse gases in it. In this instance, in addition to the same concepts present in the previous model, it is necessary for students to learn to master those of selective absorption and transparency. This new approach yields equilibrium temperatures (of the earth and atmosphere) much closer to the experimental ones, allowing us to conceptually introduce climatological modeling.

In the third and final stages, a brief introduction to climatology is presented, with emphasis on pointing out the role of human activities in CC. In addition, some misinformation techniques were analyzed with the students and the claims of old denialism were compared with those of the new one. This was also done through the proposal of activities based on interactive web-apps, like, for example, "Cranky Uncle" (https://crankyuncle.com).

5.4 Update of the Experimental Part

The experimental part is the one that has been updated the most over the years and versions (Besson et al. 2010a; Onorato et al. 2021a; Toffaletti et al. 2022, 2023). Here, too, we tried to make changes that could foster students' understanding of key concepts that were found to be more challenging.

The majority of these improved experiments utilize low-cost materials and implement a BYOD (Bring Your Own Device) strategy, partially replacing the expensive and difficult-to-obtain equipment used in earlier versions. As shown in Fig. 3, several experiments were updated as follows:

- Wien's Law: Introduced the Kanthal wire experiment, where a narrow wire is traversed by increasing current, and the spectra as a function of increasing temperature are acquired using a smartphone camera covered by a diffraction grating.
- Stefan–Boltzmann's Law and Black Body Radiation: Studied with the low-cost apparatus described in reference (Onorato et al. 2021b).
- Beer–Lambert's Law: Focused particularly on the dependence of absorption on thickness and concentration.
- IR Radiation and Selective Transparency: Refined and expanded the use of an IR camera to reinforce the concept of selective transparency.

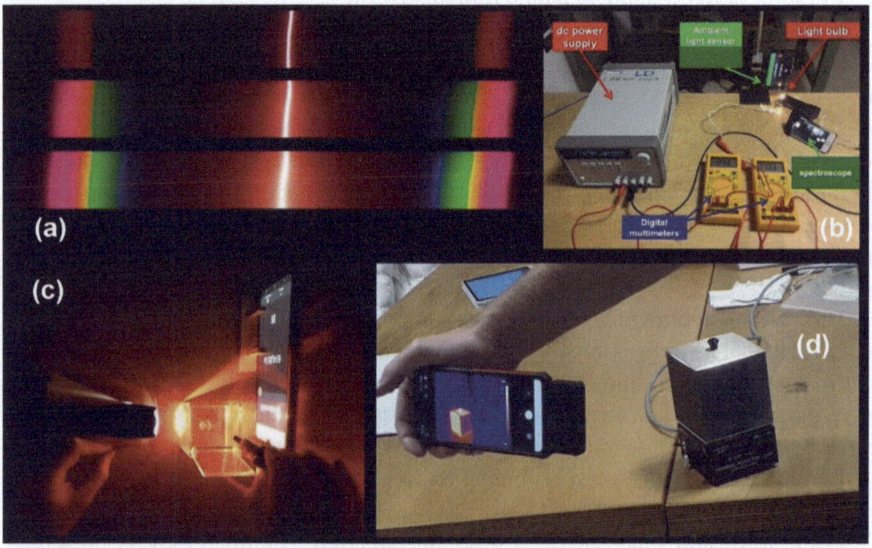

Fig. 3 The experiments: **a** the spectra of the current carrying wire at different temperatures acquired using a smartphone camera covered by a diffraction grating, **b** the low-cost apparatus to measure the blackbody radiation, **c** the experimental setup to measure the Beer–Lambert's law, **d** the IR camera employed to discuss the emissivity using the Leslie cube

These updates aim to make the experiments more accessible and cost-effective while maintaining their educational value.

6 Conclusions

In agreement with the iterative approach to design that is typical of the MER (Duit et al. 2012), the process of updating the TLS has taken into account: (1) both the results from the various research activities carried out by our group in this area; (2) "external" factors such as, for example, the re-introduction of "educazione civica" (civic education) in Italian schools; and (3) a changed awareness of teachers and students regarding GHE and CC.

The experimentation with the latest version of the TLS is still in its early stages and therefore quantitative results are not yet available. However, the increased active participation of students and their better understanding of the role of humans in the phenomenon is qualitatively appreciable.

So, we can answer to our research questions:

RQ1 The experience accumulated over the years and the results obtained from experiments with students have led to significant changes in the content and specific objectives of the sequence. The continuous effort of our research team to improve experimental techniques, combined with the increasing availability of low-cost technologies, has allowed us to modify the activities proposed to the students over the years. This process has led to an evolution of the TLS, which now includes more interactive activities and advanced technologies that facilitate student learning and promote engagement.

RQ2 In recent years, the perception of CC has changed a lot, both in public opinion and among the younger generation. Climate denialism is much less present, especially among our students, and some misconceptions, such as that of the ozone hole, seem to be outdated, as confirmed by the collected data. This evolution calls for a redesign, since teaching cannot be separated from students' ideas and mental models. Today, researchers are faced with a different pseudoscientific approach, which requires continuous adaptation of teaching content and methodologies to maintain teaching effectiveness.

The process of evolution of a teaching sequence is influenced by at least three concomitant factors: feedback from previous experiences; the changed starting conditions of the students (external factors); and an internal factor, represented by the changed conception of the phenomenon and the objectives of the teaching intervention on the part of those who carry out the planning. In the past, the main objective was to provide an incontrovertible scientific foundation for the CC, explaining the scientific basis of the GHE. Today, much of the scientific community also considers more complex factors, such as psychological factors and student engagement. This new perspective has led to the spread of the idea of considering the phenomenon in terms of all-round scientific literacy.

RQ3 While the first question can be answered in the affirmative, the measurement of the contribution of educational interventions such as these on student engagement is much more complex. As the sequence evolves toward scientific literacy objectives, there is a need for instruments that can measure these components as well. The ability to change students' engagement regarding CC is a significant objective of our work, which requires constant adaptation of evaluation methodologies to ensure the effectiveness of the proposed activities.

On the other hand, we also acknowledge that our study has some limitations. In fact, it is at present mainly focused on the Italian educational context, particularly with the recently reintroduced subject of civic education, so there may be doubts about the generalizability of the results to other educational contexts. Of course, the goals of sustainability education, particularly raising awareness of climate change, and scientific literacy about it, are relevant worldwide, being among the goals of the UN 2030 Agenda.

A second limitation is the small number of participating students and their heterogeneous prior education.

Besides addressing these limitations, we are working to implement this sequence in two new areas: that of training teachers to create learning units for civic education, and that of direct training of high school students. This kind of goal will require a further redesign to address the different levels of knowledge of the key topics and educational objectives of the discipline.

References

Besson U, De Ambrosis A, Mascheretti P (2010a) Eur J Phys 31

Besson U, Borghi L, De Ambrosis A, Mascheretti P (2010b) Int J Sci Educ 32

Center for Countering Digital Hate "Report 2024 THE NEW CLIMATE DENIAL". https://counterhate.com/research/new-climate-denial. Accessed 31 May 2024

Duit R, Gropengießer H, Kattmann U, Komorek M, Parchmann I (2012). In: Jorde D, Dillon J (eds) Science education research and practice in Europe. Sense Publishers, Rotterdam, p 13

Fiorello C, Di Mauro M, Malgieri M, Rosi T, Toffaletti S, Onorato P, Oss S (2023) In GIREP-EPEC conference 2023 proceedings—Book of extended abstracts. Slovak Physical Society, Košice

Jarrett LE (2013) Ph.D. thesis. University of Wollongong

Keller JM (2006) Ph.D. thesis. University of Arizona

Liu S-C (2021) Int J Sci Educ 43

Melbourne Centre for Behaviour Change. https://crankyuncle.com. Accessed 1 Dec 2024

Moggio L (2017) Ph.D. thesis. Università di Trento

Onorato P, Mascheretti P, De Ambrosis A (2011) Eur J Phys 32

Onorato P, Malgieri M, De Ambrosis A (2017) J Phys: Conf Ser

Onorato P, Gratton LM, Polesello M, Salmoiraghi A, Oss S (2018) Phys Educ 53

Onorato P, Malgieri M, Rosi T, Salmoiraghi A, Oss S, De Ambrosis A (2021a) G. FIS 62

Onorato P, Rosi T, Tufino E, Caprara C, Malgieri M (2021b) Eur J Phys 42

Pachauri RK, Meyer L, Core Writing Team (2014) Climate Change 2014: synthesis Report: a Report of the Intergovernmental Panel on Climate Change (IPCC)

PASCO Complete Thermal Radiation System. https://www.pasco.com/products/lab-apparatus/thermodynamics/radiation/td-8855. Accessed 31 May 2024

Psillos D, Kariotoglou P (2016) Springer
Spyrtou A, Kariotoglou P (2016) In: Psillos D, Kariotoglou P. Springer
Stocker T (2013) Climate Change 2013. The Physical Science Basis: Working Group I Contribution to the Fifth Assessment Report of the Intergovernmental Panel on Climate Change (Intergovernmental Panel on Climate Change)
Tasquier G (2013) G. Fis. 54
Toffaletti S, Di Mauro M, Rosi T, Malgieri M, Onorato P (2022) Educ Sci 12
Toffaletti S, Di Mauro M, Malgieri M, Rosi T, Tufino E, Onorato P, Oss S (2023) Am J Phys 91
University of Colorado Boulder. https://phet.colorado.edu. Accessed 31 May 2024
White R, Gunstone R (1992) Probing understanding. Routledge, London

Laboratory in Teaching Physics

Recent Perspectives in Physics Education Through the Adoption of Hardware and Software Technologies in the Laboratory

Eugenio Tufino, Tommaso Rosi, Pasquale Onorato, and Stefano Oss

Abstract The increasing use of technology in physics education opens up new perspectives for teaching and learning. Among them, hardware devices such as iOLabs, smartphones, and augmented reality devices, together with software tools such as Python, are redefining the experimental approach in teaching. iOLabs and smartphones, integrating advanced sensors and an intuitive interface, promote active scientific inquiry, where students play an active role in testing hypotheses or models, making decisions, and finding solutions on their own. We will show how the ability of such devices to compute and graph the Fourier transform of sensor signals in real time can be used to explore various physical phenomena in new ways. Augmented reality, implemented through visors, smartphones, and other devices, provides direct immersion in physical principles, transforming abstraction into concrete experience. Finally, Jupyter Notebooks in Python are emerging as user-friendly tools for data processing, allowing students to develop computational skills through exercises and applications directly related to experiments. In this paper, we present examples of teaching modules, some of which follow the ISLE approach, designed for students and teachers in the last years of high school and in introductory university courses.

E. Tufino (✉) · P. Onorato · S. Oss
Department of Physics, University of Trento, Povo (Trento), Italy
e-mail: eugenio.tufino@unitn.it

P. Onorato
e-mail: pasquale.onorato@unitn.it

S. Oss
e-mail: stefano.oss@unitn.it

T. Rosi
Level up S.r.l., Trento, Italy
e-mail: tommaso.rosi@leveluptrento.com

I. Testa et al. (eds.), *Connecting Physics Education Research and Practice*,
Challenges in Physics Education, https://doi.org/10.1007/978-3-031-86609-8_7

1 Introduction

In this paper, we examine the integration of various digital technologies, which have become increasingly common and affordable in recent years, into physics education. The evidence in support of active learning strategies in both theoretical and laboratory settings has become overwhelmingly compelling (Hake 1998; Von Korff et al. 2016; Wieman 2014), highlighting the importance of active learning in which students are continuously required to apply and process their knowledge in a variety of ways. However, laboratory activities are not inherently conducive to active engagement. Recent research on laboratory practice has also highlighted the limitations of overly structured and confirmatory laboratory setups (Smith and Holmes 2021).

Inquiry-based approaches have emerged that are inherently based on active learning, but also allow for student-led inquiry activities with appropriate scaffolding. These approaches allow students to learn physics by engaging in the practice of physics and accommodate different levels of inquiry. Notable examples for high school or introductory college courses include SCALE-UP (Beichner et al. 2007) and Modeling Instruction (Brewe 2008) and the Investigative Science Learning Environment (ISLE) (Etkina and Van Heuleven 2007). Documents such as the American Association of Physics Teachers (AAPT) recommendations for laboratory work (American Association of Physics Teachers 2014) and frameworks for laboratory work such as (Zwickl 2014) provide valuable guidance for defining learning objectives for different laboratory contexts. Digital technologies, both software and hardware, can serve as tools to achieve these learning objectives, such as data collection and analysis, visualization, and design of experiments. They offer the potential for more open-ended laboratory experiences, and active engagement and encourage an investigative approach, as opposed to traditional cookbook-style labs. The hardware technologies we explore include activities using sensor-based devices, smartphones, thermal cameras, and augmented reality headsets. On the software side, we discuss the introduction of computational elements through Jupyter Notebooks in Python.

In the following sections, we first provide a general overview of these technologies and their applications, followed by a more detailed discussion of how our research group has implemented them.

2 Digital Technologies as Educational and Supporting Tools

This section outlines some of the digital technologies that have had a significant impact on physics education, highlighting their role as teaching and support tools. We use the distinction between hardware and software technologies, although it is clear that the boundary between them is subtle and not always clearly defined. These technologies, both hardware and software, enable the use of Multiple External Representations (MERs) (Treagust 2017), which facilitate different approaches to teaching and learning physics topics. In addition, the effectiveness of these representations is enhanced by the fact that most of these devices provide real-time data (Brasell 1987). Together, these digital tools can be used to achieve the learning goals (American Association of Physics Teachers 2014) such as analyzing and visualizing data, modeling physical phenomena, designing experiments, and developing technical skills.

Digital technologies can play a crucial role in bridging physical phenomena and their mathematical representations in physics education. As highlighted by Euler et al. (2023), these tools support both mathematization and visualization processes, by linking hands-on experimentation to formal theoretical models. Visualization tools, in particular, can connect physical phenomena directly to mathematical formalism, as well as link idealized models to formal representations. By offering students visual, hands-on ways to engage with abstract physics concepts, these technologies make formal ideas more accessible and relatable to students' experiences, enhancing both mathematization and visualization in physics learning.

2.1 Hardware Technologies

Sensor Devices: The advent of affordable and versatile sensor devices such as Arduino, smartphones, and devices like Interactive Online Lab System (iOLab) (iOLab 2024) has transformed data collection in physics experiments. These devices are equipped with a range of sensors capable of measuring various physical parameters such as acceleration, magnetic field strength, and temperature. Their integration into the classroom allows students to perform a wide range of experiments in all areas of introductory physics, from mechanics to optics to electromagnetism, facilitating hands-on learning and the development of experimental skills. For example, Kuhn and Vogt's (2022) compilation of over 70 examples from a decade of publications in "The Physics Teacher" (Kuhn and Vogt 2022) highlights the diverse applications of smartphones in educational settings. Similarly, a recent resource letter by Monteiro and Martí Monteiro and Martí (2022) provides a comprehensive overview of the improvements in student learning facilitated by these devices. Another useful resource is (O'Brien 2021). Hundreds of experiments have been developed using Arduino, significantly expanding the repertoire of activities published in leading physics education journals.

In some initiatives we have used the (iOLab), a multi-use sensor device developed by the Physics Education Group at the University of Illinois and distributed by Macmillan. The iOLab contains a number of high-resolution sensors that are capable of measuring a wide range of physical phenomena, including acceleration, force, light intensity, magnetic fields, and voltage. It connects wirelessly to computers and allows real-time analysis through its comprehensive graphical interface. This interface supports simultaneous data analysis from multiple sensors and provides tools for detailed statistical evaluation, such as calculating Fourier transforms and analyzing area under curves.

Examples of research on the sensors devices' effectiveness in improving conceptual understanding and positive attitudes to learning include (Becker et al. 2018; Hochberg et al. 2020; Mazzella and Testa 2016).

Infrared (IR) cameras: In recent years, there has been a notable introduction of thermal imaging cameras (with sensors sensitive to IR radiation in the 8–15 micron range) into the market, followed by their adoption in education research and school laboratories, at relatively low costs and with highly efficient management software. The literature on this subject has similarly expanded, making it increasingly impractical to exclude this technology from educational programs dedicated to a wide range of physical science applications, particularly in thermodynamics (Vollmer and Möllmann 2017; Haglund et al. 2022). The potential uses are numerous, and in our laboratory, the IR thermal camera is often employed as a vital supplementary tool, enriching observations made with traditional methods.

Augmented Reality (AR): Over the past two decades, the increasing availability and ease of use of Virtual and Augmented Reality technologies has placed them at the forefront of Science and Physics Education research (Ibáñez 2018; Akçayır and Akçayır 2017). Various augmented reality setups, based on smartphones, tablets, 3D cameras, projectors, glasses, and viewers, have seen a lowering of their costs and the introduction of new user-friendly software. Our research group has been actively engaged in several projects in this direction, utilizing AR in laboratory settings and science communication events, which we will discuss in this article.

2.2 Software Technologies

Simulations: Digital simulations provide an interactive platform for exploring physics concepts without the need for physical equipment. Simulations such as PhET, Algodoo, and VPython create virtual environments where students can manipulate variables, observe results, and engage in exploratory learning.

Video analysis tools: Tracker (Brown 2008) and similar video analysis software allow students to analyze the kinematic and dynamic properties of moving objects. It also facilitates investigations in optics (Rosi et al. 2016).

Python and Jupyter Notebooks: In the rapidly evolving technological world, computational skills have emerged as a critical asset in various fields, including physics, so there's a growing consensus for their inclusion in the physics curriculum, going beyond traditional tools such as spreadsheets and software packages. The inclusion of computational elements in the physics curriculum has taken various forms, from including computational exercises in existing lecture courses to incorporating computational tasks in laboratory settings. This movement toward computation is further supported by specific guidelines from the AAPT (Behringer 2024) and the Partnership for Integration of Computation into Undergraduate Physics (PICUP) (Caballero and Chonacky 2019). The inclusion of computational thinking as one of the science practices in the Next Generation Science Standards (NGSS) (NGSS Lead States 2013) has encouraged a synergy with undergraduate physics courses. Jupyter Notebooks provides an interactive computing environment that easily integrates Python code, equations, visualizations, and text. This allows students to visualize physical systems, perform computational modeling, and analyze data in a comprehensive way.

3 Case Studies: Implementing Digital Technologies in Physics Experiments

3.1 iOLab Device and ISLE Approach

We explored the adoption of the ISLE approach with the innovative use of iOLab digital devices across different physics topics in high school physics. This approach was applied specifically to the unit on Newton's laws for 11th-grade students and to the DC circuits unit for 12th-grade students.

Our learning objectives were three, based on AAPT guidelines (American Association of Physics Teachers 2014), NGSS practices (NGSS Lead States 2013), and Bloom's Revised Taxonomy (Anderson et al. 2001):

- **Conceptual knowledge:** Students will understand and apply the physical concepts behind the experiments.
- **Technical Skills:** Students will be competent in the use of the iOLab apparatus and the Phyphox smartphone application (RWTH Aachen University 2024) for all aspects of experimental work.
- **ISLE Process:** Students will engage in inquiry-based activities that reflect the ISLE process.

It is important to note that in this intervention, theoretical lessons were conducted by the classroom teacher, while laboratory activities involving the iOLab system were facilitated by an external expert (one of the authors). Using the iOLab equipment, we initiated an investigation into dynamic friction. Students observed the acceleration curve generated by the accelerometer as the device slid across their desks (see Fig. 1). They were then asked to test their hypotheses by using the force sensor in conjunction with the accelerometer. This process concluded with an **application experiment** (Etkina et al. 2006) to estimate the coefficient of dynamic friction using two different methods, providing a hands-on experience that actively engaged students in the physical investigation process.

Empirical data collected during these activities highlights significant improvements in students' conceptual understanding and technical skills. Pre- and post-intervention self-assessment surveys indicate significant gains across multiple scientific abilities, as evidenced by effect sizes. For example, students reported enhanced confidence in designing experiments, analyzing data, and evaluating experimental results. Final test assessments further supported these results, showing improvements in three identified dimensions: ISLE process, conceptual knowledge, and technical

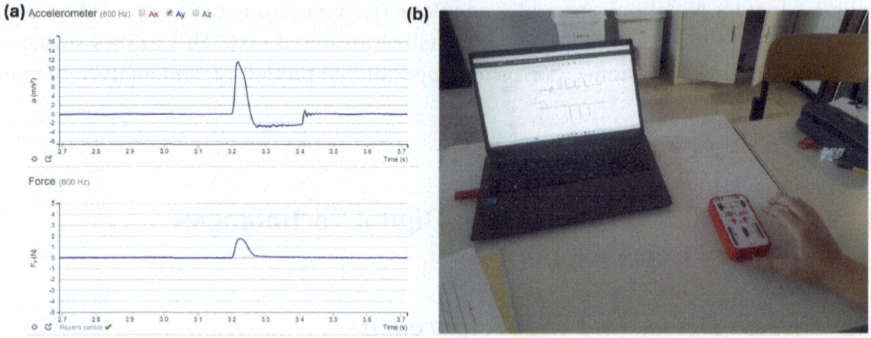

Fig. 1 **a** Students measure force and acceleration using iOLab accelerometer and force sensors, giving the device a push to slide it across a desk. **b** The setup shows the iOLab device connected to a computer

skills. Results indicate a majority of students achieving proficiency levels categorized as "needs improvement" to "near mastery," with stronger outcomes observed in the 12th-grade classes due to extended instructional time. Additionally, feedback emphasized the intuitive design of the iOLab setup and its effectiveness in bridging theoretical models with experimental practice, as discussed in (Tufino et al. 2025a; Tufino 2024).

In this module activity, students constructed a horizontal Atwood machine by measuring acceleration with the wheel sensor and tension with the force sensor (see Fig. 2). The activity highlighted the discrepancies between the theoretical model and the actual data, primarily due to the frictional effects of the string on the table edge. Students revised their assumptions and later introduced a pulley system to reduce friction, better aligning their experimental data with theoretical predictions. This iterative process mirrors the ISLE approach by actively engaging students in hypothesis revision and validation.

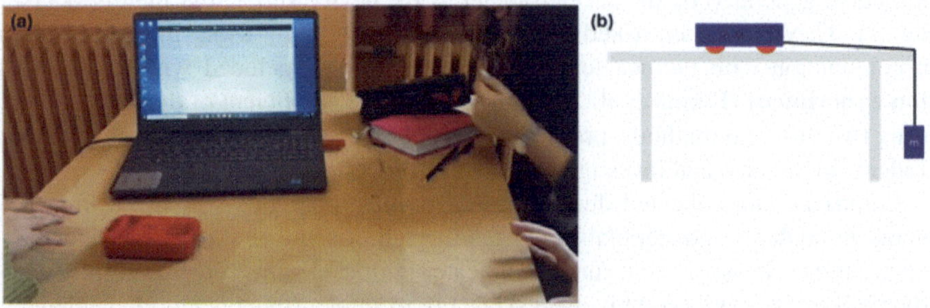

Fig. 2 **a** Experimental setup by a student group, showcasing the initial attempt to measure force and acceleration (without a pulley). **b** Schematic representation of the Atwood machine problem as typically presented in textbooks

In the DC circuits module, students explored the behavior of incandescent bulbs and LEDs in simple circuits. This ISLE-based module began with observational experiments in which students experimented with lighting both an LED and an incandescent bulb separately, discovering how the orientation of the components affected the circuit's functionality. Testing experiments required students to evaluate their initial explanations (Etkina and Brookes 2019), followed by group discussions addressing common misconceptions documented in physics education research (McDermott and Shaffer 1992). These discussions, facilitated by whiteboard activities (Fig. 3), supported peer learning and promoted conceptual understanding, consistent with ISLE practices.

Students investigated Ohm's law and I-V characteristics by constructing circuits to measure potential differences and calculate current for LEDs and incandescent bulbs. These activities emphasized key concepts such as the asymmetrical nature of the I-V curve for LEDs, highlighting their directional behavior in contrast to the non-directional characteristics of incandescent bulbs. Subsequently, using the iOLab system and its Digital-to-Analog Converter (DAC) to vary the applied voltage, students constructed the I-V curve of an LED and identified its distinct threshold voltage, linking this observation to their earlier qualitative investigations. This activity not only reinforced their understanding of electrical phenomena but also emphasized data analysis and technical skills development, highlighting the effectiveness of integrating iOLab and ISLE methodologies (Tufino 2024).

Students quickly became familiar with the iOLab system, and their feedback highlighted the importance of group work and hands-on activities in developing engagement and conceptual understanding. The limited time frame and separation of laboratory activities from theoretical lessons limited the full realization of the potential of the ISLE approach.

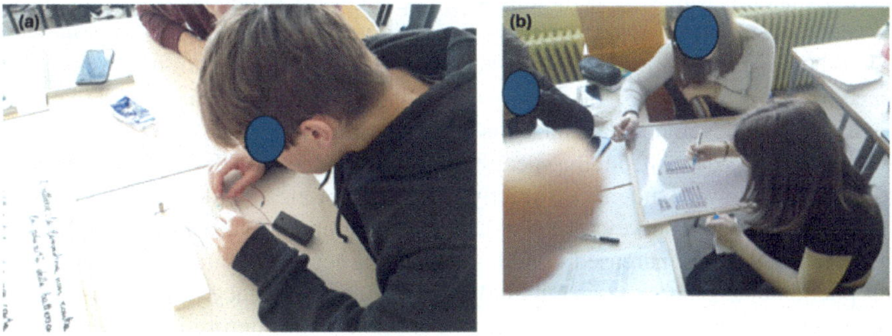

Fig. 3 **a** Students working on a circuit set up with an incandescent bulb. **b** Students collaborating around whiteboards, discussing conceptual questions related to circuit behavior

4 Smartphone and FFT Amplitude Spectrum Plot

In this module for 12th-grade students, we explored the potential of smartphones to provide real-time amplitude spectrum analysis via Fast Fourier Transform (FFT), introducing this visual representation as an additional form of multiple representations to enhance student inquiry and understanding of physical phenomena (Tufino 2024). The module was structured to guide students from conceptual understanding to practical application, integrating hands-on activities and technology-based tools to promote engagement and inquiry.

The module began with an introductory activity using a Jupyter Notebook in Google Colab. Students were presented with an unknown signal composed of multiple frequencies and were tasked with identifying the individual frequencies manually by analyzing the waveform. After this initial challenge, they were introduced to FFT as a tool to simplify the process, providing a clear demonstration of how this technique can decompose complex signals into their frequency components. This activity emphasized FFT's practical application in analyzing periodic phenomena.

The next part of the module focused on two experimental activities that utilize the Phyphox app for real-time spectrogram analysis. In the first activity, students explored Dual-Tone Multi-Frequency (DTMF) signaling, a telecommunication system that encodes key presses with unique tone pairs. Working in pairs, one student generated DTMF tones using an online tone generator while the other used Phyphox to analyze the spectrogram and decode the corresponding keys. This exercise linked theoretical concepts, such as wave superposition, to real-world applications, fostering both analytical and technical skills.

The second activity investigated the physics of sound in musical instruments, specifically the relationship between the length of wind instruments and the pitch of the sound they produce. Starting with examples like piccolos and bassoons, students studied standing waves in air columns and applied this understanding to estimate the speed of sound. Using a straw and a smartphone microphone, students generated harmonics by blowing into the straw and analyzed the spectrogram to calculate the speed of sound. This hands-on activity not only reinforced concepts related to sound waves but also developed data analysis and modeling skills.

The visual feedback from the spectrograms, combined with the intuitive interface of Phyphox, allowed students to directly connect theoretical models to experimental observations. A post-survey revealed that students found the activities enriching and enjoyable, reporting increased confidence in using technology to investigate physical phenomena. Final assessments indicated strong proficiency in both conceptual understanding and technical application, particularly in interpreting experimental data and visual representations. The use of FFT amplitude spectrum analysis is not limited to the sound-related activities in this module but can also be applied to other physics topics as a tool for investigating complex phenomena and enriching the high school physics curriculum. An example of another proposed lab activity is described in (Tufino et al. 2023).

4.1 Augmented and Virtual Reality

Setup AR

Over the past few years, our group has developed and tested several hardware/software setups for using AR in the physics laboratory: in this section, we will briefly present two setups. The first, known as "The Augmented Laboratory" (The Augmented Laboratory Project 2024; Rosi et al. 2017), exploits a 3D camera and projector. This system can create 3D reconstructions of the physical world, captured by the 3D camera. These reconstructions not only enhance the visual representation of experiments but also allow for detailed post-experiment analysis, providing a powerful tool for inquiry-based learning and cultivating critical thinking skills. These reconstructions can be viewed in real time as data are captured or explored later through interactive playbacks. The system is capable of tracking, i.e., capturing the positions in time, of multiple 3D objects (Rosi et al. 2017) and projecting the relevant data directly onto and around the objects used in real physics experiments-what we call an "augmented experiment." This integration of real-time data projection into physical experiments bridges the gap between abstract theoretical models and tangible observations, encouraging deeper conceptual understanding. In Fig. 4, two frames of a video recorded during an augmented experiment with this setup can be seen.

This setup also introduced the potential for new interactions between computer simulations and the real world, which we call "augmented simulations." An example of augmented simulation is an augmented ripple tank (see Fig. 5). This setup allows various wave propagation experiments to be performed using physical objects to represent wave sources, barriers, lenses, and areas with varying wave velocities. By enabling students to manipulate physical objects while observing the corresponding wave behaviors in real time, this approach promotes active learning and encourages hypothesis-driven experimentation. The software can recognize and differentiate these physical elements that are introduced into the projection area. As a result,

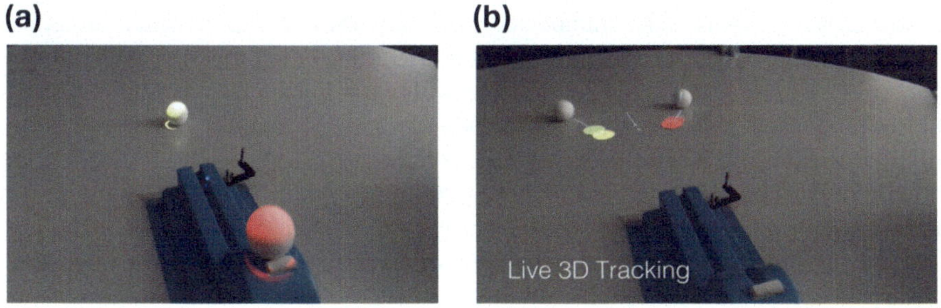

Fig. 4 **a** Real-time tracking of the position of two colliding spheres is shown; **b** the results of the tracking can be recorded and reviewed, even adding additional information such as vectors and graphs

Fig. 5 An "augmented simulation" of a ripple tank

this tool facilitates an intuitive and clear discussion of phenomena such as reflection, refraction, diffraction, Doppler effect, and sonic boom. This hands-on exploration aligns with the theory of embodied cognition, which suggests that learning is grounded in sensorimotor systems, making abstract concepts more accessible through physical interaction.

The second main setup we worked on was one based on virtual reality (VR) and AR headsets (AnReAL Project 2024; Rosi et al. 2017). The idea behind their use is similar to that of the previous setup: to be able to obtain quantitative measurements using the setup and visualize the results of these measurements directly in the physical world (e.g., tracking the controllers and visualizing their 3D movement in real time using the viewers, see Fig. 6) and to be able to combine computer simulations and real experiments. This system forms the basis of the Augmented Reality Active Learning (AnReAL) framework (AnReAL Project 2024), which integrates AR tools into physics education to create active learning environments. The AnReAL framework encourages students to engage with experiments by predicting, testing, and refining their hypotheses in an interactive, immersive environment. Additionally, the use of VR/AR headsets enables collaborative learning by projecting the experimental data onto external monitors, allowing groups of students to discuss and analyze results together. This collaborative approach mirrors professional scientific practices and supports the development of teamwork and communication skills alongside technical competencies.

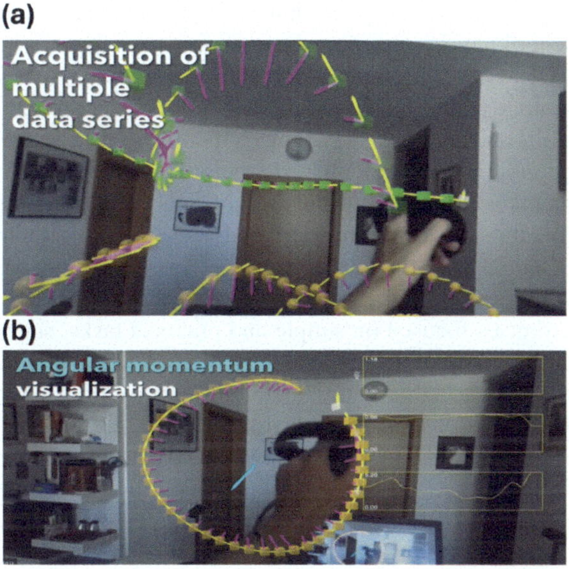

Fig. 6 Real-time 6-degree-of-freedom tracking of a controller, viewed through an AR viewer and simultaneously on the computer monitor. At the top you can see the acquisition of multiple data sets (represented by different colors) and at the bottom the display of graphs and vectors such as angular momentum (in cyan)

Student Engagement and Preliminary Insights from AR/VR Implementation

Preliminary testing with students showed that AR/VR setups are a promising addition to traditional tools in physics courses, although further research is needed to reach stronger conclusions. As part of the Augmented Reality Active Learning (AnReAL) framework (AnReAL Project 2024), these setups were evaluated for their ability to enhance conceptual understanding and engagement. Students noted that the real-time data acquisition and visualization capabilities simplified the process of analyzing experiments, particularly in tracking motion with six degrees of freedom. For example, the visualization of angular momentum vectors in 3D was reported to clarify abstract concepts, in line with the aims of embodied cognition theories.

Students also appreciated the interactive nature of the AR/VR setups, which allowed them to dynamically test and refine their hypotheses. Although some students required initial guidance to familiarize themselves with the technology, the overall feedback highlighted the accessibility and effectiveness of these tools (Rosi et al. 2022).

4.2 Jupyter Notebooks in Python

To introduce computational elements, we used Python and Jupyter Notebooks. This is exemplified by two different implementations: the introduction of Python for data analysis in a first-year physics laboratory course at the university level, and the application of computational concepts and practices through an FFT module in high school physics (see Sect. 4).

The integration of Python and Jupyter Notebooks into the first-year physics laboratory course was designed as a pilot project to enhance students' computational skills and promote active engagement with data analysis tasks. This initiative builds upon an already established laboratory structure, incorporating computational elements without significantly altering the course content or increasing the workload for students. Using Jupyter Notebooks, the project aims to familiarize students with essential tools for data manipulation, visualization, and data analysis, aligning these skills with the experimental practices already present in the curriculum. The initial exercises focused on simple and practical tasks, such as plotting graphs and calculating averages, to build students' confidence and competence before progressing to more advanced activities, including data handling, uncertainty analysis, least-squares fitting, and visualization (Tufino et al. 2024, 2025b).

The computational learning goals we developed were inspired by the recommendations of the American Association of Physics Teachers (AAPT) (Behringer 2024), initiatives such as PICUP (Partnership for Integration of Computation into Undergraduate Physics), and other works in Physics Education Research (PER). For a more extensive discussion of the relevant literature, see (Tufino et al. 2024). The computational learning goals are centered around two key aspects:

1. **Implement**: enabling students to develop programming skills for manipulating, analyzing, and visualizing data;
2. **Communicate**: teaching students to effectively document their work, create clear and comprehensive graphs, and write well-structured code.

The course structure was adapted to integrate these computational goals without significantly altering the existing curriculum. For instance, traditional tools such as SciDavis were replaced by Python programming and laboratory exercises were modified to include computational components such as linear regression, error analysis, and graphical data representation. Students worked with Jupyter Notebooks containing real-world examples, such as analyzing atmospheric CO_2 data (see Fig. 7), simulating physical systems, and interpreting experimental results.

Pedagogical Reflections and Challenges

The Jupyter Notebooks were designed and implemented using a gradual and engaging approach, with multiple phases aimed at lowering the barrier to programming while emphasizing the context of physics. Initially, students were introduced to simple tasks focused on analyzing experimental data they had already collected, such as plotting graphs and calculating averages. Gradually, the complexity of the exercises increased, incorporating tasks such as fitting data, performing uncertainty analysis, and simulating physical phenomena. This progressive structure ensured that students developed their programming skills along with their understanding of physics, developing a smooth integration of computational tools into laboratory experience (Tufino et al. 2024). To ensure that all students, regardless of their initial programming experience, could fully engage with the computational tools, the curriculum

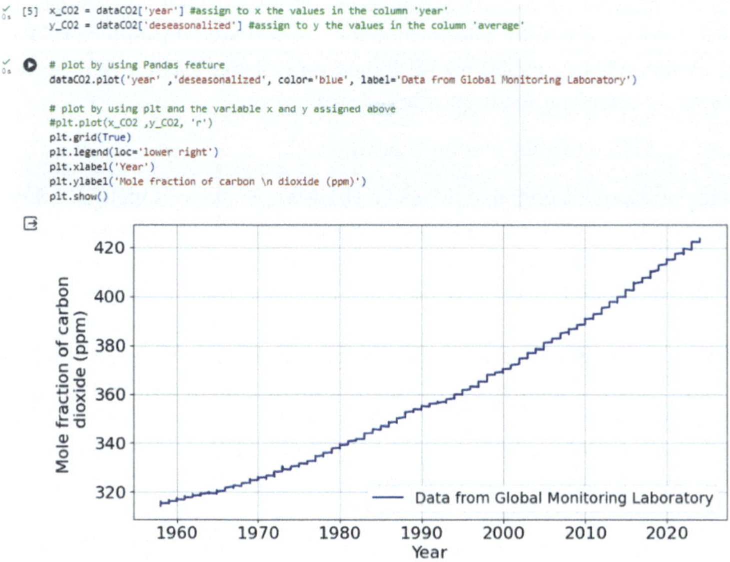

Fig. 7 This figure shows a portion of our Jupyter notebook that integrates text, code, and graphs to analyze CO_2 concentration trends importing CO_2 data directly from the Mauna Loa Observatory website

provided scaffolded learning opportunities and collaborative group work. Working in groups allowed students to share knowledge and support each other, encouraging a collaborative learning environment that facilitated active participation.

The evaluation process provided valuable information on the effectiveness of this implementation. At the beginning of the course, students completed a questionnaire assessing their attitudes toward computation and their initial level of programming experience. At the end of the course, a follow-up questionnaire was administered, including open and structured questions, to evaluate their perceptions, learning outcomes, and areas of difficulty. In addition, the laboratory notebooks of the students were analyzed to assess their ability to integrate computational tools into their data analysis and interpretation. The findings reported in (Tufino et al. 2024) indicated significant improvements in students' confidence and competence in using computational methods, alongside a deeper understanding of the underlying physics.

The experience gained from this intervention in computation highlights the feasibility and effectiveness of integrating computational tools into university-level physics education. By aligning these tools with broader educational objectives, such as fostering inquiry-based learning and developing professional skills, this approach provides a valuable framework that can be adapted by other institutions. Additionally, the integration of Python and Jupyter Notebooks into the laboratory curriculum demonstrates how computational methods can enhance students' understanding of experimental physics and prepare them for future academic and professional challenges. In parallel to the university-level initiative, simple computational elements were introduced to high school students, in particular through an FFT module described earlier. This provided useful indications of the feasibility of introducing computational elements at the high school level.

The Role of LLMs and Computation Education

The advent of Large Language Models (LLMs), such as OpenAI's ChatGPT, Anthropic's Claude, and Google's Gemini, has introduced significant challenges and opportunities to the teaching and learning of computational skills. These tools can assist students in generating code, debugging, and exploring programming concepts. However, their availability raises important pedagogical questions. How should educators adapt computational curricula to ensure that students develop a deep understanding of coding and data analysis, rather than becoming excessively dependent on automated solutions?

Our proposed module, which integrates Python and Jupyter Notebooks into the curriculum, provides a foundation for addressing some of these challenges, though further adaptations are necessary. As Yeadon et al. suggest, LLMs can serve as powerful complementary tools in coding education, provided that students are first equipped with foundational skills to critically evaluate and interpret the outputs of these models (Yeadon et al. 2024). By emphasizing group work and inquiry-based learning, we aim to encourage a deep understanding of coding practices, enabling students to use AI tools as valuable resources without undermining essential core skills.

5 Conclusions

Our exploration of the integration of new technologies into physics teaching, as demonstrated by case studies in high school and introductory university courses, underscores the enhanced engagement and understanding facilitated by these tools. Although space constraints limit our discussion to a selection of technologies, it's clear that these technologies are redefining teaching methods, particularly when used within inquiry-based approaches such as the Investigative Science Learning Environment (ISLE). However, in the case of the ISLE-based intervention, certain practical constraints, such as the limited number of hours dedicated to the module and the separation of laboratory activities (facilitated by the external researcher) from theoretical lessons (conducted by the classroom teacher), impacted the full realization of the ISLE approach's potential. The examples presented illustrate the potential of tools such as iOLab, Phyphox applications, AR visors, and Jupyter Notebooks to drive innovation in teaching practices and improve student engagement in experimental physics. More generally, for these technologies to be effective in the classroom, teachers need not only a structured and active framework that integrates theoretical and practical components but also adequate technical skills and confidence to use these tools effectively. Supporting teachers in this is crucial to ensure that students can fully benefit from the educational opportunities offered by digital technologies.

References

Akçayır M, Akçayır G (2017) Advantages and challenges associated with augmented reality for education: a systematic review of the literature. Educ Res Rev 20:1–11

American Association of Physics Teachers (2014) AAPT recommendations for the undergraduate physics laboratory curriculum. Am Assoc Phys Teach

Anderson LW, Krathwohl DR, Bloom BS (2001) A taxonomy for learning, teaching, and assessing: a revision of bloom's taxonomy of educational objectives. Longman

AnReAL Project (2024) AnReAL project webpage. https://www.anrealproject.org/. Accessed in Mar 2024

Becker S, Klein P, Gößling A, Kuhn J (2019) Using mobile devices to augment inquiry-based learning processes with multiple representations. J Sci Educ Tech

Behringer E, Burciaga J, Dietz D, Gavrin A, Kozminski J, Migenes V, Schroeder D (2016) AAPT recommendations for computational physics in the undergraduate physics curriculum AAPT. https://www.aapt.org/resources/upload/aaptuctfcompphysreportfinal.pdf. Accessed in Mar 2024

Beichner RJ, Saul JM, Abbott DS, Morse JJ, Deardorff DL, Allain RJ, Bonham SW, Dancy MH, Risley JS (2007) Student-centered activities for large enrollment undergraduate programs (SCALE-UP). Physics

Brasell H (1987) The effect of real-time laboratory graphing on learning graphic representations of distance and velocity. J Res Sci Teach

Brewe E (2008) Modeling theory applied: modeling instruction in introductory physics. Am J Phys 76:1155–1160

Brown D (2008) Video analysis and modeling in physics education. J Res Sci Teach

Caballero MD, Chonacky MN, Engelhardt L et al (2019) PICUP: a community of teachers integrating computation into undergraduate physics courses. Phys Teach 57:397

Etkina E, Murthy S, Zou X (2006) Using introductory labs to engage students in experimental design. Am J Phys 74:979–986

Etkina E, Brookes D T, Planinsic G (2019) Investigative science learning environment. Morgan & Claypool Publ

Etkina E, Van Heuleven A (2007) Investigative science learning environment—a science process approach to learning physics. Am Assoc Phys Teach 1

Euler E, Solvang L, Gregorcic B, Haglund J (2023) Visualization and mathematization: how digital tools provide access to formal physics ideas. In: Taşar MF, Heron PRL (eds) The international handbook of physics education research: special topics. AIP Publishing. pp 21-1–21-28

Haglund J, Jeppsson F, Schönborn KJ (2022) Thermal cameras in science education. Springer, Cham

Hake RR (1998) Interactive-engagement versus traditional methods: a six-thousand-student survey of mechanics test data for introductory physics courses. Am J Phys 66:64–74

Hochberg K, Becker S, Louis M, Klein P, Kuhn J (2020) Using smartphones as experimental tools-a follow-up: cognitive effects by video analysis and reduction of cognitive load by multiple representations. J Sci Educ Tech 29:303–317

Ibáñez MB, Delgado-Kloos C (2018) Augmented reality for STEM learning: a systematic review. Comp Educ 123:109–123

iOLab (2024) iOLab device homepage. https://www.iolab.science. Accessed in Mar 2024

Kuhn J, Vogt P (2022) Smartphones as mobile minilabs in physics. Springer

Mazzella A, Testa I (2016) An investigation into the effectiveness of smartphone experiments on students' conceptual knowledge about acceleration. Phys Educ 51:055010

McDermott LC, Shaffer PS (1992) Research as a guide for curriculum development: an example from introductory electricity. Am J Phys 60

Monteiro M, Martí AC (2022) Resource letter mds-1: mobile devices and sensors for physics teaching. Am J Phys 90:328–343

NGSS Lead States (2013) Next generation science standards: for states, by States. National Academies Press, DC

O'Brien DJ (2021) A guide for incorporating e-teaching of physics in a post-COVID world. Am J Phys 89:403–412

Rosi T, Malgieri M, Onorato P, Oss S (2016) What are we looking at when we say magenta? Quantitative measurements of RGB and CMYK colours with a homemade spectrophotometer. Eur J Phys 37:065301

Rosi T, Onorato P, Oss S (2017) Multiple object, three-dimensional motion tracking using the xbox kinect sensor. Eur J Phys 38:065003

Rosi T, Baldassi M, Melucci A, Onorato P, Oss S (2022) Augmented reality active learning (AnReAL) activities for teaching/learning motion concepts. J Phys: Conf Ser 2297:012005

RWTH Aachen University (2024) Phyphox application. https://phyphox.org/. Accessed in Mar 2024

Smith EM, Holmes NG (2021) Best practice for instructional labs. Nat Phys 17:662–663

The Augmented Laboratory Project (2024) The augmented laboratory project webpage. https://www.augmentedlab.org/. Accessed in Mar 2024

Treagust DF, Duit R, Fischer HE (2017) Multiple representations in physics education. Springer, Cham

Tufino E (2024) Rethinking laboratory activities with digital technologies and developments in physics education. PhD thesis, University of Trento, Trento, Italy

Tufino E, Gratton L, Oss S (2023) Analyzing light signals and AC frequencies with FFT. Phys Educ 2:3500130

Tufino E, Oss S, Alemani M (2024) Integrating python data analysis in an existing introductory laboratory course. Eur J Phys. https://doi.org/10.1088/1361-6404/ad4fcc

Tufino E, Onorato P, Oss S (2025a) Exploring active learning in physics with ISLE-based modules in high school. J Phys: Conf Ser 2950:012021. https://doi.org/10.1088/1742-6596/2950/1/01202

Tufino E, Oss S, Alemani M (2025b) Using Jupyter Notebooks to foster computational skills and professional practice in an introductory physics lab course. J Phys: Conf Ser 2950:012022. https://doi.org/10.1088/1742-6596/2950/1/012022

Vollmer M, Möllmann KP (2017) Infrared thermal imaging: fundamentals, research and applications. Wiley, NY

Von Korff J, Archibeque B, Gomez KA, Heckendorf T, McKagan SB, Sayre EC, Schenk EW, Shepherd C, Sorell L (2016) Secondary analysis of teaching methods in introductory physics: a 50k-student study. Am J Phys 84:969–974

Wieman CE (2014) Large-scale comparison of science teaching methods sends clear message. Proc Natl Acad Sci USA 111:8319–8320

Yeadon W, Peach A, Testrow C (2024) A comparison of human, GPT-3.5, and GPT-4 performance in a university-level coding course. Sci Rep 14:23285

Zwickl BM, Finkelstein N, Lewandowski HJ (2014) Incorporating learning goals about modeling into an upper-division physics laboratory experiment. Am J Phys 82:876–882

Laboratory Experience of Optical Reflection in a Circular Cavity

Veruska Lamberti, Roberto De Luca, Francesco Saverio Tortoriello, and Immacolata D'Acunto

Abstract The educational proposal focuses on the investigation of the recursive properties of the phenomenon of light reflection in a circular cavity within the framework of geometric optics. The research targets primary and secondary school students and is based on an experiential approach: the reflection trajectory within a circular cavity is reproduced and studied. By this experimental activity primary school students may gain familiarity with the geometric and recursive properties of light reflection using ruler and compass constructions. Middle school students can enhance these skills by analyzing and verifying the incommensurability properties of real numbers, leading to the interdisciplinary construction of regular polygons from component triangles. Finally, high school students can extend this study to trigonometric periodic functions, can analyze the application of the induction principle underlying the recursive property of reflection within the mirror, and can generate computer images of trajectories.

V. Lamberti (✉) · R. De Luca
Department of Physics 'E.R. Caianiello', University of Salerno, Fisciano, SA, Italy
e-mail: velamberti@unisa.it

R. De Luca
e-mail: rdeluca@unisa.it

F. S. Tortoriello
Department of Mathematics, University of Salerno, Fisciano, SA, Italy
e-mail: fstortoriello@unisa.it

I. D'Acunto
Department of Human, Philosophical, and Educational Sciences, University of Salerno, Fisciano, SA, Italy
e-mail: idacunto@unisa.it

I. Testa et al. (eds.), *Connecting Physics Education Research and Practice*,
Challenges in Physics Education, https://doi.org/10.1007/978-3-031-86609-8_8

1 Introduction

Research on light has a long and rich history, dating back to antiquity. Various scientists, philosophers, and thinkers from different eras have contributed to its understanding (Watson 2016). Democritus (Baldes 1978) laid the embryonic theory of corpuscular nature. Witelo, in his treatise *Perspectiva* (Witelo 2025), developed a complex analysis of light and vision, addressing phenomena like reflection and refraction, the geometry of light rays and color perception, combining scientific knowledge with philosophical and metaphysical interpretations. Roberto Grossateste traced the origin of all matter back to a point of light (Grosseteste and Panti 2007) (considered by some as a precursor to the Big Bang theory), linking the concept of energy-force to it, albeit always within a metaphysical and philosophical framework. Later thinkers such as Leonardo Da Vinci (Kemp 2007), Johannes Kepler (Donahue 2000), inspired by Witelo (Kepler 2025), and Isaac Newton contributed to the birth of geometric optics (Newton 1952). Christiaan Huygens (Andriesse 2005), on the other hand, developed the wave theory of light. Augustin Jean Fresnel (Huygens 1900) used the concept of transverse waves to explain phenomena such as diffraction and interference. James Clerk Maxwell (Glazebrook 1896; Maxwell 1873) unified the equations of electromagnetism, demonstrating that light is an electromagnetic wave. In the twentieth century, Albert Einstein had the insightful idea of using the invariance of the speed of light to unify space and time into a continuum called "spacetime". This concept overturned our classical perception of our surroundings, paving the way for a new paradigm and giving rise to a new awareness of the physics of the universe (Kaku 2004). In this historical panorama we propose a cross-sectional study on reflection of light rays in the realm of geometric optics. Providing students with the opportunity to carry out simple experiments with light will allow them to develop critical skills, refine competencies, and transfer knowledge from one field to another (Dewey 2013; Schon 2013). The interdisciplinary character of natural phenomena will enable students to recognize the contribution of each discipline to the understanding of the physical phenomenon (Morin 2014). In the context of teaching and for a broader and more comprehensive treatment, it is advisable to introduce the dual nature of light from primary school onwards, in a context that is not strictly mathematical, since this is obviously impossible, but rather a purely debate-driven and experimentaldiscovery of the intriguing and hard-to-explain characteristics of light (see Sect. 3.1). A strict adherence to the results of science demands not to reduce the problem of light propagation to the corpuscular nature alone, which, though more immediately accessible for mental visualization (Michelene et al. 1994), would not provide students with a complete framework for understanding light phenomena and the knowledge acquired so far (Stefanidou and Skordoulis 2014). Considering that the minds of primary and secondary school students are fertile ground for misconceptions (Galili and Hazan 2000; Blizak et al. 2009) and misunderstandings, it is essential to avoid fostering incomplete mental frameworks and guide reasoning appropriately. Such distorted logic could lead to dysfunctional constructs and sometimes entirely erroneous ones (Allen 2019). The literature appears to focus on studying and analyzing the efficiency of an education system that tends to implement a stratification of knowledge, which is stacked and accumulated based on an easily intuitive and solid characteristic of light (the corpuscular one), rather than debating the methods and timing of introducing the more elusive but equally important wave characteristic. Indeed, the latter is introduced in school curricula as a "surprise" at more advanced levels of education (secondary school, university). In short, this subject is not completely studied starting for primary school. However, with relief, we observe that the introduction of modern physics is part of the experimental science curriculum in primary schools in Greece (Voudoukis and Kalkanis 2007; Testa et al. 2020). For this reason, our proposal is directed at primary and secondary school students, hoping that a well-structured laboratory context can contribute, as much as possible, to improving the understanding of the phenomenon of light propagation and to provoke curiosity and questions (even if unresolved), as well as to habituate young minds to philosophical and historical discussions (Garritz 2013; Henriksen et al. 2018). The laboratory proposal illustrated here must therefore be an integral part of a comprehensive experiential program, which should include multiple laboratory activities and must be paired, as previously stated, with experiences on the wave nature of light (Tóth et al. 2024).

2 The Didactic Proposals

The educational proposal (Fig. 1) is structured according to Bruner's spiral curriculum (Bruner 1966), revisiting the theme of light reflection in a circular cavity across three educational levels, each adapted to students' cognitive abilities. Following criteria of representation, economy, and effectiveness, the approach develops through three methods of representation that enable the construction of a progressively more complex curriculum for studying and understanding the phenomenon. The proposal thus develops gradually, adapting to students' different ages and cognitive capacities: for primary school, an '*active representation*' involving practical actions to achieve a result is used (see sect. 3), (see Tables 1 and 2). The construction of summary and symbolic images applies the '*iconic representation*' suitable for lower secondary school students (see sect. 4.1), (see Tables 1 and 2). Finally, the '*symbolic representation*' method used in the proposal for upper secondary school is based on the use of logical propositions of a symbolic system (see sect. 4.2), (see Tables 1 and 2). The entire process is characterized by a multidisciplinary and interdisciplinary approach.

The educational proposal is purely constructivist in primary school, enriched with cognitive elements in lower secondary school, and adds connectivist strategies in upper secondary school. The IBSE (National Research Council et al. 2000) approach is common across the three educational levels, with different TLS (Teaching Learning Sequence) for each level. The choice is based on the intention to stimulate students' critical thinking, their ability to ask questions, to design and conduct investigations, create arguments, build models, and communicate results, with the aim of deepening understanding using logic and evidence concerning the natural world (Crawford 2014). IBSE is an approach through which students learn science in a

Table 1 Description of teaching proposals and related methods of representation in accordance with Bruner's theory

School level	Duration	Prerequisites	Activity description	Methods of representation
Primary school	3–4 h	Basic concepts of light and shadows	Practical activity: representation of the reflection phenomenon using a ruler and compass	Active representation (direct experience)
Lower secondary school	4–5 hours	Basic geometry concepts	Iconic representation of reflection in a circular cavity through the construction of regular polygons	Iconic representation (images and graphs)
Upper secondary school	8–9 hours	Basic knowledge of trigonometry	Study of the recursive properties of reflection using trigonometric functions and calculation software	Symbolic representation (symbolic systems and software)

Table 2 Domains involved, teaching strategies and methodologies applicable. ***Domains:*** 1. Epistemic, 2. Procedural, 3. Conceptual, 4. Social, 5. Affective. ***Teaching Strategies and Methodologies:*** *Student-centered approaches*: (a) Autonomous, (b) Collaborative, (c) Discovery-based. *Active teaching methods*: (d) Guided discussion, (e) PBL (Problem-Based Learning), (f) PjBL (Project-Based Learning), (g) Brainstorming . *Inductive and deductive approaches*: (h) Inductive, (i) deductive. *Tool-based approaches*: (j) Gamification, (k) Use of digital technologies. *Metacognitive strategies*: (l) Think aloud, (m) Planning and monitoring, (n) Self-assessment and reflection, (o) Problem solving. *Motivational strategies*: (p) Storytelling *Concept-based approaches*: (q) Principles and theories, (r) Concept transfer. *Evaluation strategies*: (s) Quiz, (t) Authentic assessment, (u) formative assessment, (v) process assessment

School level	Domains	Teaching strategies and methodologies
Primary school	1, 2, 4, 5	a, c, d, e, g, h, i, j, l, p, q, s, u, v
Lower secondary school	2, 3, 4	b, e, f, h, i, m, n, o, q, r, t, u, v
Upper secondary school	2, 3, 4	c, e, f, h, i, k, m, n, o, q, r, t, u, v

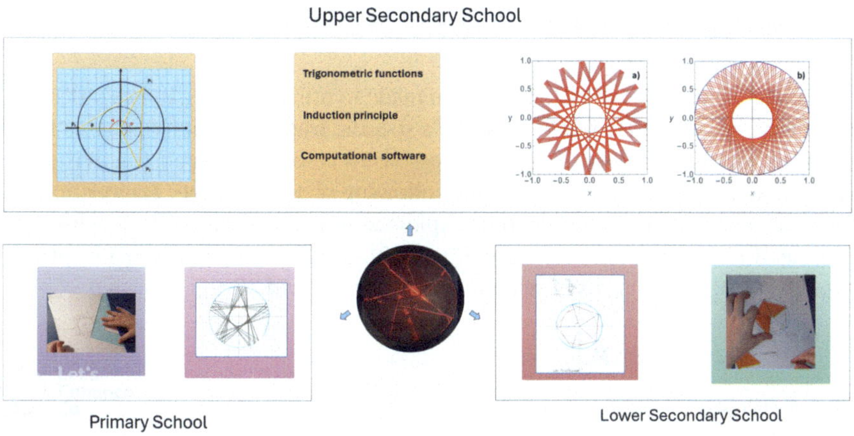

Fig. 1 The educational proposal for the three levels of education

completely natural way, as it aligns with their way of exploring the world (Crawford 2014) and adopting it with PSTs (Pre-Service Teachers) (Fazio 2014) during pre-role training enhances their capabilities across different teaching domains (Strat et al. 2024), just as it does for students's formation. Furtak et al. (2012) found that IBSE had a greater impact on learning when teachers connected students' arguments to prior knowledge and academic concepts, which justifies placing our proposal within a constructivist spiral framework and recommends its application at all three educational levels presented here. The tables below summarizes the main characteristics of the three educational proposals, including duration, students' prior knowledge, strategies used, methods, and domains covered. The TLS proposed for the different educational levels are illustrated in the respective sections. Currently, the proposal that has been tested is the one for primary school.

Tables 1 and 2 summarize the main characteristics of the teaching proposal tailored to various levels of education and list the teaching methodologies and strategies that can be used. The teaching strategies and methodologies will be referenced in the following text using the associated letter, showed in Table 2, through the symbols (x) or (x, y, ...).

Experimental Apparatus. The apparatus used to carry out the experiment is shown in Fig. 1. It consists of a circular base with a diameter of 30 cm, on the perimeter of which a mirror has been mounted. The dimensions of this mirror are slightly smaller than the length of the circumference, leaving a small opening with a maximum size of 3 cm. Using a high-coherence and low-dispersion laser, a light beam is introduced into the mirror through this opening and the multiple reflections of the light ray will thus create different geometric figures.

3 Teaching Proposal for Primary School

For primary school, the decision has been made to utilize Rodger Bybee's BSCS 5E model (Bybee et al. 2015), as it provides a specific framework in five phases for the design and implementation of scientific lessons. It is rooted in a constructivist teaching approach. The model comprises the following phases: engage, explore, explain, elaborate, and evaluate, which will be explained in the following. The teaching proposal, as illustrated below, was tested in a fifth-grade class (final year) of primary school with 18 students, 13 of whom were boys, all aged 10.

3.1 Engage

To spark interest, stimulate curiosity, and encourage questions, the teacher presents an age-appropriate video (suitable for 10-year-old students in their final year of primary school). The video centers on the scientific debate about the nature of light and concludes by acknowledging the contributions of both Newton's particle theory and Huygens' wave theory (*q, d*). This approach allows students, who hold a naive view of the phenomenon, to understand that scientific progress often arises from the comparison of different ideas and that the nature of light, in time, will prove to be much more complex than previously imagined. Thus, students can begin to familiarize themselves with the scientific environment in which they will be operating, be encouraged to formulate questions, and try to find answers.

The laboratory activity begins with the story (*p,d*) of "*Mrs. Light*" in which we have chosen to personify light (and therefore photons), giving her a name and appearance (Fig. 2a) and making her the protagonist of an exciting story where the circular cavity becomes a mirror game called "*The Billiard Mirror*" (Fig. 2b). Through the story, fundamental properties of light are introduced, such as the absence of mass, energy transport, speed, and, naturally, its dual nature. These striking characteristics of light often lead to a brainstorming session (*g, d*) (right after watching the video), where students, emotionally engaged, are ready to respond to various questions, including three main ones (see Fig. 2c). The discussion was animated and full of pedagogical insights (for instance, revealing misconceptions to address).

The story ends with a request to solve Mrs. Light's problem of escaping the cavity (*e, j*). At this stage, the main objective is not to explain the profound dual nature of light but to spark curiosity and start building the mental foundations upon which to stack knowledge about the wave nature of light and prepare them for the laboratory activity. Experimental activity about wave light property can be introduced in later activities and explored in future stages of their education.

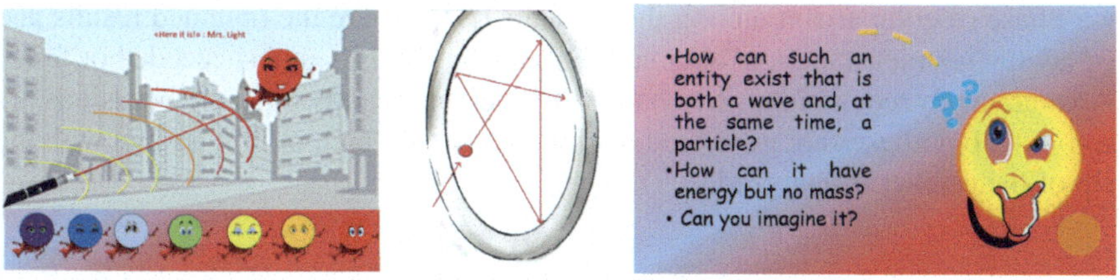

(a) (b) (c)

Fig. 2 **a** Mrs Light and the Photon family. **b** The Billiard Mirror. **c** Guiding questions for brainstorming

3.2 Explore

The explore phase is therefore based on the necessity to help Mrs. Light escape from the Billiard Mirror. Thus, the students draw their circumference on grid paper and open a small door at the intersection from the x-axis and the circumference to allow the light to enter (*a,c*). They will use the following sequence to discover if Mrs. Light will be able to escape for the billiard. The ruler and compass construction is suggested in (De Luca 2010) and reported below. Let assume that the point particle enters the outer circumference of center O at P_0, as shown in Fig. 3a, at an angle θ with respect to the horizontal. If we were just to consider the particle's trajectory inside the circumference, as shown in Fig. 3a the path of the light is obtained in the following simple way: first draw the segment P_0-P_1; then draw segments P_{n-1}-P_n all tangent to the circumference of radius d equal to the distance between the point O and the segment P_0-P_1. This procedure gives the starred patterns in Fig. 3b. The proposed drawing activity allows students to construct the recursive (*h*) path within the circumference themselves, making them aware of how a property (n-th trajectory) can be a function of the same preceding property ((n-1)-th trajectory). This is followed

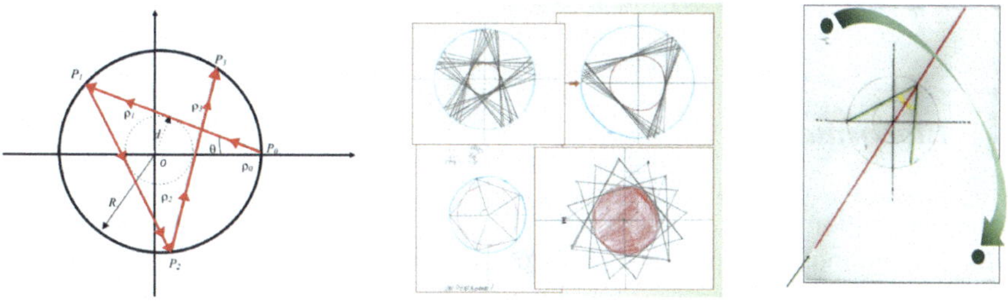

(a) (b) (c)

Fig. 3 **a** Trajectory of a point particle bouncing elastically within a circumference of radius R and center O. **b** Same drawing obtained by 10-year-old students. In the case of the pentagon, Mrs. Light managed to exit. **c** A 10-year-old student's artifact to verify the congruence of the angles of incidence and reflection

by a phase of comparison among the students (*l*), to share the obtained results and display the geometric figures obtained. Cases where Mrs. Light succeeded or failed to exit are analyzed, and the reasons for this happening only on some occasions are explored. Hypotheses are generated, and the geometric path of light drawn with the help of the teacher is analyzed (*d, i*).

3.3 Explain

In this phase, the knowledge acquired by the students does not yet have a defined structure; it appears allocated and inserted into pre-existing structures in an unstable and sometimes inconsistent manner. The first logical connections needed to create support for the entire scaffold are beginning to form with the help of discussions and comparisons held in the previous phase.

Through classroom experimentation, the presence of students' belief that the internal circumference with radius d, as seen in Fig. 3a, used to geometrically construct the path of light, is genuinely present, i.e. has a real physical consistence, has been observed. In order to dispel this belief, it is important to point out, also by showing the equipment, that the circumference of radius d does not exist physically and that its use is only aimed at the geometric construction of the path of light inside the mirror. At this point, it is possible to introduce students to the laws of reflection, using a flat mirror initially to highlight the congruence of the angles of incidence and reflection. One can then proceed to analyze the case of a circular cavity, which can be easily verified through a second and quick graphical construction as illustrated below. Using a transparent sheet placed on the previous squared one, students will reproduce Mrs. Light's entry path into the circular cavity. The geometric reconstruction will end at the first reflection (*a*); they will then draw a colored line connecting the center of the circle with the first point of incidence. Then they will be invited to fold the transparent sheet (*c*) along this colored line to verify, through the overlay of the incident ray and the reflected ray, the congruence of the obtained angles (*d*). This process is schematically illustrated in Fig. 3c.

3.4 Elaborate

In the elaboration phase, students extend and apply their acquired knowledge and skills through more complex and applied activities. The equipment is shown to the students to allow them to observe and verify what they have learned about the laws of reflection. The different reflection paths and, consequently, the various geometric shapes that can be obtained are shown to the students (*j*) (Fig. 4).

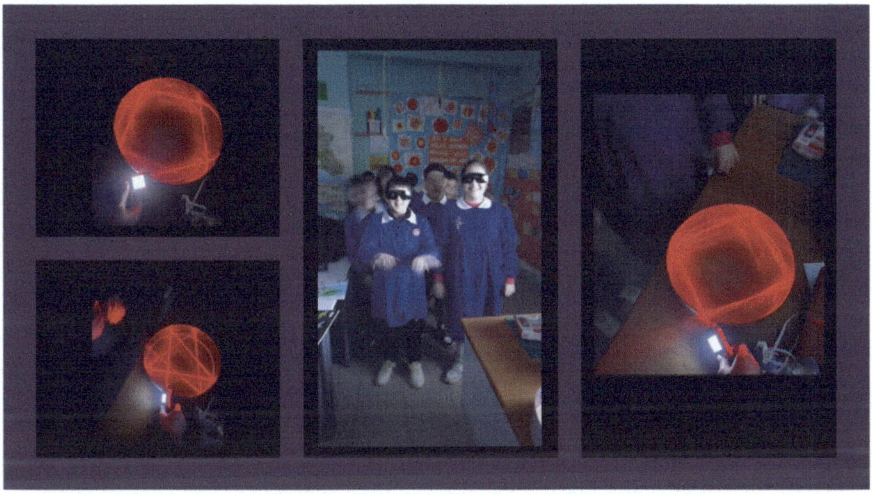

Fig. 4 In the elaborate phase the equipment is shown to the students

3.5 Evaluate

During the learning assessment session, a questionnaire with multiple-choice and open-ended responses was administered. After each question, the intended learning objectives that the question aimed to verify are included. Below are the questions we consider most significant for pedagogical and research purposes and the most highlighted results.

Assessment questionnaire

1. Now, do you have a clearer idea of Light? Light therefore is: (a) A particle, (b) A wave, (c) Both[1]

   The question aims to investigate whether students have developed an initial awareness of the complexity of light nature, recognizing that it can behave in different ways (particle and wave nature). It also seeks to stimulate curiosity and reflection on the different scientific models used to interpret natural phenomena, preparing them for more in-depth learning in future experiences.

2. From your drawing, did Mrs. Light manage to exit the billiard she entered? (a) Yes, (b) No
   The question aims to investigate whether students have understood the iteration and the result of the geometric construction of the light path within the "Billiard Mirror."

[1] Obviously, in the limit of geometric optics, in which light is considered to be represented by luminous rays, the correct answer to this question cannot be achieved (see 3.1).

3. Explain, in your opinion, why she managed to exit or why she could not.
   The question aims to investigate the students' ability to explain, based on their observations, the necessary conditions for light to exit the cavity, consolidating their understanding of the geometric construction and the properties of reflection.
4. Looking at your drawing, which polygon did Mrs. Light draw?
   The question aims to investigate the students' ability to correctly identify the geometric shapes produced by the construction of the light path, connecting them to the concepts of iterative reflection.
5. Which of these geometric figures never be drawn? You can give more than one answer.
   (a) Circle, (b) Circumference, (c) Square, (d) Triangle, (e) Hexagon, (f) Pentagon.
   The question aims to investigate the students' understanding of the relationship between the properties of light reflection and the geometric shapes that can or cannot be produced due to the geometric construction.
6. Why can not Mrs. Light enter the unknown area[2] in any way? (a) Because there is no door, (b) Due to geometric construction, (c) Both.
   The question aims to ascertain whether students understand the geometric constraints and the conditions of the problem, particularly the role of the geometric construction in determining the path of light.
7. Was it difficult for you to draw Mrs. Light's path? If yes, why? If no, why?
   The question aims to investigate students' perception of the difficulty of the activity, in order to understand their level of familiarity with iterative geometric constructions and their practical skills.
8. What new words have you learned today?
   The question aims to verify the acquisition of the scientific and geometric vocabulary introduced during the activity, such as "particle," "photons," "circumference," and terms related to the phenomenon of reflection.

Discussion on the questionnaire and results analysis

1. 95% of students responded: *both*, and the 5% responded *wave*. The dual nature of light has been accepted. No student answered *corpuscle*, although the corpuscular property of light is employed in the geometric construction of the trajectory.
2. 85% of students answered yes, and 16% answered no.
3. From the obtained responses, it is understood that students have grasped the iterative nature of the trajectory construction operation.
4. 50% of the students answered with a *triangle*, and the other 50% with a *star*, even though the drawings representing a triangle were only one in total. This percentage distribution of responses can be understood by considering that a star is composed of multiple triangles.
5. 83% of the students answered correctly by choosing the options a) and b) among the various choices. However, 57% of these students also added reproducible

[2] Author's Note: Circumference of radius d.

figures such as the square, hexagon, and pentagon, even though these were reproduced during the explain phase. Therefore, there is a conceptual misalignment between what was experienced on paper and what was found by experimental observation. 17% chose options c, d and e.
6. 55% of the students answer *due to geometric construction*, 33% *because there is no door*, and 12% did not respond.
7. 12% of students answer *yes*, because of the number of iterations needed to get the light out of the mirror.
8. 70% of the students respond *corpuscle*, with small percentages of students adding the words *photons, straight trajectory, billiard, laser, circumference*.

The teaching proposal aims to develop a series of essential educational objectives in students. Through the exploration activity, it fosters autonomy in problem-solving and critical analysis skills, encouraging the active discovery of geometric properties. It promotes inductive and deductive reasoning, strengthening logical and systematic thinking. Students will acquire technical skills such as geometric drawing and the ability to construct visual models to understand complex phenomena. Peer comparison helps improve social skills, such as effective communication. Moreover, the activity supports the development of metacognitive thinking, encouraging students to reflect on learning processes and monitor their progress. Storytelling and the motivational context enhance engagement and stimulate curiosity (Fuchs and Corni 2023), while the application of principles and theories strengthens the ability to transfer abstract concepts to concrete contexts.

4 Teaching Proposal for Secondary School

From the responses to question 3. and 7. (Sect. 3.5) it is evident that students had become familiar with the recursive properties of reflection in the circular cavity, iterating them until they achieved the goal of freeing Mrs. Light from continuous bouncing within the circular mirror. While this phenomenon was observed in the case of the graphic activity proposed to young students, it may not be universally applicable. The geometric and algebraic conditions for this to occur arise only when the angle α in the Fig. 5b is a rational multiple of the angle 2π, according to: $2k\pi = N\alpha$ (De Luca 2023), where k and N are integers. On the other hand , for $\alpha = 2\pi\gamma$, where γ is an irrational number, escape from the circular cavity will not be possible, if the entrance hole is a simple point. The experiment with Mrs. Light will thus be suitable for geometric studies. The following educational proposals for lower and upper secondary school is still in the illustrative phase and has not yet been tested in the classroom.

4.1 Proposal for Lower Secondary School (students Aged 11–13 Yrs)

The proposal for lower secondary school aims to offer students the opportunity to study the geometric properties underlying the multiple reflections of a laser beam in a circular cavity, with a focus on understanding the properties of regular polygons, developing skills in using geometric tools, connecting geometric properties to abstract mathematics, stimulating deductive and inductive reasoning, exploring the concept of incommensurability, and fostering collaborative work and metacognitive strategies.

The students are introduced to the tools and perform the construction with ruler and compass proposed to primary school students. In a collaborative environment, the students investigate possible cases in which the iterative procedure of the light beam's path results in a finite number of iterations (*e*). This occurs only when the light beam traces regular figures along its path. The students are then asked to analyze the geometric properties of such polygons and relate them to the angle θ with which the beam enters the circular cavity (*m, n, o*). This procedure will lead them to discover that the figure in question is a regular polygon made up of a different number of isosceles triangles with side R, depending on the polygon obtained. Students are asked to draw such a triangle on a cardboard using a compass and ruler, and repeat the drawing N times[3]. By cutting and assembling N triangles, various regular polygons with N sides can be obtained, all inscribed in a circle with radius R. From the geometric solution of the triangles, the validity of the theorem on the sum of the interior angles of a convex polygon can be deduced (*i*), reflecting on the concept of polygons inscribed in a circle, and finally verifying that regular polygons with N sides can be decomposed into N equilateral triangles (*q, r*). Students are asked to construct equilateral triangles whose vertex angles have a strictly random magnitude, but which have radius R, and to verify the possibility of constructing further inscribed regular polygons (*f, m, n, o*). The students are led (*h*) to realize that this happens only when the ratio between 2π and α belongs to $\mathbb{Q}$, and that when this is not the case, the value of such a ratio is incommensurable. The assessment follows an authenticity criterion.

The teaching proposal is based on an integrated approach that combines discovery-based learning, collaborative work, and methodologies such as Project-Based Learning (PjBL). Inductive and deductive strategies, along with the necessary metacognitive practices of monitoring, planning, self-assessment, and reflection, will develop critical and transferable skills in students. The goal is to promote active and conscious learning, where students not only acquire theoretical knowledge about light but are also able to apply it in practical contexts, such as reflection on flat and curved mirrors. In this way, students will be introduced to tackling and solving real-world challenges, both independently and collaboratively.

---

[3] To construct the triangle on the cardboard, the value of the vertex angle α must be calculated through integer division, see Fig. 5a.

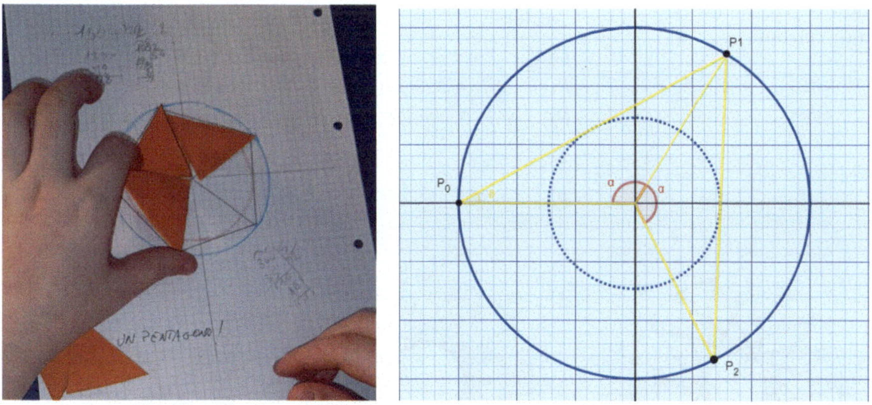

(a) (b)

Fig. 5 **a** A student's artifact to reconstruct a pentagon. **b** Graphical construction of the trajectory of light

4.2 Proposal for Upper Secondary School (students Aged 17–18 Yrs)

The entire learning process unfolds in an experiential and collaborative IBSE-STEAM environment, structured and guided by a framework modeled on Kolb's learning cycle (Kolb 2014). This approach is particularly effective for methodologies such as Project-Based Learning (PBL), where the focus is on concrete activities and abstraction processes culminating in the completion of a project (PjBL). The model provides a framework for designing lessons that integrate hands-on activities, guided reflections, and theoretical applications, enabling students to achieve a deep and transferable understanding while actively participating in the learning process. The four phases of the cycle: *concrete experience, reflective observation, abstract conceptualization, and active experimentation* can repeat continuously, creating a dynamic learning process that adapts to new operational needs.

For upper secondary school, we propose expanding the teaching approach presented in the previous sections, particularly the observation of the experimental setup in operation and the ruler-and-compass reconstruction (*c*) (*concrete experience*), by incorporating activities that include the use of symbolic formalism (*abstract conceptualization)* typical of mathematics and computer science *(active experimentation)*.

Each point of incidence can be associated with a vector in the complex plane using trigonometric or exponential formalism; let us assume that the laser enters the cavity at point P_0. Therefore, we can associate a vector with a magnitude of R and an initial phase equal to π to the light path. It should be noted that the new incident point is identified by the clockwise rotation of the previous vector by an angle of magnitude α, as shown in Fig. 1b. The graphical construction (Reflective Observation) of the

Table 3 Trigonometric representation

Vector	Trigonometric notation
P_0	$R\,[\cos(\pi - 0\alpha) + i\,\sin(\pi - 0\alpha)]$
P_1	$R\,[\cos(\pi - 1\alpha) + i\,\sin(\pi - 1\alpha)]$
P_2	$R\,[\cos(\pi - 2\alpha) + i\,\sin(\pi - 2\alpha)]$
$\vdots$	$\vdots$
P_k	$R\,[\cos(\pi - k\alpha) + i\,\sin(\pi - k\alpha)]$

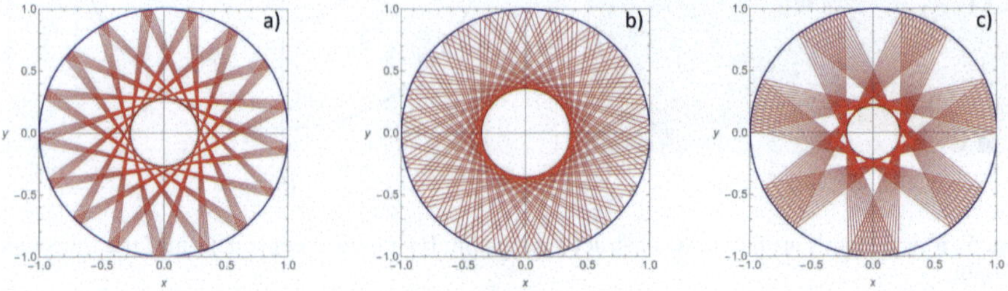

Fig. 6 Reflections of a light ray inside a circular cavity with radius $R = 1$, in arbitrary units. In all panels, 100 reflections are shown. The initial inclination of the light ray is $\theta_0 = \pi/(6\sqrt{2})$ in panel (a), $\theta_0 = \pi/(8\sqrt{2})$ in panel (b), and $\theta_0 = \pi/(10\sqrt{2})$ in panel (c)

isosceles triangle $P_n P_{n+1} O$ with a vertex angle α serves as a visual aid to help students identify, step by step, the arguments of the trigonometric functions represented by the vector in the complex plane *(i, q)*. From the image, it can be observed that the argument of the trigonometric function represented in the complex number P_0 coincides with the phase π. For the subsequent points (P_1, P_2, ...), since the rotation is clockwise, the argument can be obtained by subtracting from the initial phase a number of angles α equal to the number of reflections considered. It is recommended to organize such information in tabular form (see Table 3) so that students can highlight the recursive nature of the law and apply the principle of induction to it *(abstract conceptualization)*. Students will generalize the mathematical expression of the reflection law inside the circle *(h)*, without necessarily having to prove it by induction.[4]

The path of light inside the cavity, once formalized in terms of a mathematical algorithm *(r)*, can be appropriately translated into computer code through a coding operation in the symbolic language *(k, m, n, o)* of the chosen software *(active experimentation)*. In the presented images (Fig. 6), the software used is Mathematica® (De Luca and Lamberti 2024). The final assessment can follow criteria of authenticity *(t)* and it can be preceded by formative and process assessments *(u,v)*.

[4] The proof of the general nature of the recursive law can be obtained by identifying the base of induction (the vector associated with P_0), assuming the validity of the law for point P_n and demonstrating the validity for point P_{n+1}, and deducing that it is valid for every n. This demonstration can be achieved through the use of trigonometric formulas.

The educational experience as proposed can be used in an extracurricular activity or in an advanced course, although it can be adapted (by removing the formalization in programming language) into an experiential activity within the curriculum for fourth-year upper secondary school students.

The proposed teaching approach, through the development of observation and analysis skills, promotes critical and creative thinking to build knowledge independently. Students engaged in solving complex problems, such as creating code using a programming language, will learn to work in teams and acquire decision-making skills. The development of research skills will enable them to apply knowledge interdisciplinarily, leveraging analytical and synthetic abilities. They will apply the principles of reflection to real-world cases, strengthening logical reasoning; they will develop technical computing skills and deepen algebraic concepts, understanding how complex processes can be automated. In this task, they will utilize and develop computational thinking. They will therefore need to rely on their abilities and competencies to transfer scientific and mathematical knowledge into the field of computer science. They will necessarily reinforce their capacity for abstraction and acquire a systemic perspective.

5 Conclusion

Based on the experimentation carried out, the importance of an interdisciplinary, historical, and epistemological approach emerges as crucial to engaging students in a deeper understanding of natural phenomena. From the positive results of the experimentation, in terms of awareness of the complexity of the phenomenon and the degree of student engagement, it has become evident, in our opinion, that there is a need to introduce the dual nature of light as early as primary school to promote a more comprehensive understanding and stimulate critical and philosophical thinking in young minds. In this respect, the integration of IBSE principles ensures active student engagement, the development of critical thinking, and logical reasoning skills at all levels.

The teaching proposal for primary school aims to develop a deep and critical understanding of optical reflection in students by employing a series of practical activities that encourage interaction and reasoning. The geometric reconstruction of the light's path in a circular cavity, through successive iterations, enables the introduction of the principles of reflection and the geometry of mirrors within a context that is easily comprehensible and, above all, enjoyable for the students.

The activities carried out in the laboratory were fundamental in allowing learners to experiment with and concretely verify the laws of reflection using simple yet effective tools. Constructing the light's path through the manipulation of geometric shapes and analyzing the obtained results fostered the development of mathematical and scientific skills while also stimulating curiosity and reflection on the behavior of light in different contexts.

The results from the evaluation questionnaire indicate that students managed to understand the iterative nature of constructing the light's path, recognizing the importance of successive reflections and correctly associating the geometric shapes generated by the propagation of light rays. The near-universal acceptance of the dual nature of light, though not fully explored on this occasion, represents, in our view, an important initial step toward introducing a more complete and scientific perspective. Additionally, the analysis of responses to more specific questions highlighted that some concepts, such as the relationship between reflection and geometry, were well assimilated, while others require deeper exploration to address gaps in understanding, particularly regarding polygons inscribable in a circle. Some responses indicate a not full understanding regarding the geometric shapes light can trace inside the circular cavity. It may therefore be useful to delve deeper into the link between successive reflections and the generated figures, clarifying the concept of symmetry and specifying how different entrance angles affect the light path.

The use of technical language, such as "particle" and "photons," contributed to enriching the students' scientific vocabulary, bringing them closer to a more precise and formal language. Although there are still points to clarify and further develop, the experimental approach and active student engagement were decisive factors in achieving the educational objectives.

Regarding the dual nature of light, although 95% of students "accepted" this property, a deeper understanding of these concepts, especially in such a simplified context, requires further exploration and subsequent phases of laboratory and pedagogical experimentation. Nevertheless, we believe it is useful to propose activities at this educational level.

The educational proposal for lower secondary school, as structured, offers a unique opportunity for "knowledge transfer" across different disciplines: starting with the study of physics concepts, it progresses to geometric construction and, finally, to algebraic analysis. This transition from physics to geometry is fundamental for understanding how physical concepts can be described using geometric language. Once geometric comprehension is achieved, the analytical and experiential framework can be expanded to algebra, exploring relationships between angular ratios and rational and irrational numbers.

The educational activity, as described, stimulates metacognition and the development of cognitive abilities that regulate and support the transition from concrete thinking (based on observation) to abstract thinking. Through discussions on incommensurability, students can understand how mathematics addresses physical phenomena that cannot be described using simple numbers but require more advanced concepts, such as those pertaining to the algebra of rational and irrational numbers.

The classroom experimentation of this educational proposal will help evaluate how students respond to the interdisciplinary challenge and how effectively they can transfer concepts learned from one domain to another, fostering a vision of knowledge not as isolated compartments but as a continuous flow between different disciplines.

The proposal for upper secondary school could offer numerous advantages, but it also presents challenges and areas requiring further development, also considering that it has not yet been tested.

The IBSE-STEAM approach allows for the integration of knowledge in physics, mathematics, geometry, and computer science, fostering the development of transversal skills in solving complex problems.

Direct experience, implemented through practical and collaborative activities aimed at the construction of graphs and the definition of analytical-computational processes, can represent another significant advantage. The collaborative approach can enhance motivation, self-esteem, and cohesion among group members.

The software simulation of light paths using dedicated programming tools such as Mathematica® or similar programs offers students the opportunity to develop technical skills in computer science, a fundamental ability in a modern professional and academic context.

Nevertheless, there are also challenges to address. One of the main challenges is the theoretical preparation required, which may represent an obstacle for some students, particularly those who struggle with abstract concepts or using advanced software. In this context, the integration of interdisciplinary and innovative methodologies requires adequate preparation and continuous teacher training to face the innovative teaching challenges and promote high-quality scientific education.

In-service training, in particular, must be strengthened to ensure that teachers possess the necessary tools to renew school curricula and meet the demands of a modern and scientifically updated education system (Michelini 2023).

Moreover, the proposed activities require a critical selection based on the structural availability of software and hardware resources, as well as a careful analysis of time requirements. Although we have estimated these elements, they cannot be validated without experimentation. Another critical aspect could concern adapting the proposal to students with varying levels of preparation. In such cases, it is advisable to offer the section dedicated to software programming to highly motivated hight school students.

The future investigations of this proposal will allow for closer observation of its didactic valence. For example,it will be important to verify whether the combination of experiments, theory, and computer applications can lead to a deep and lasting understanding. Specifically, the level of student engagement and motivation can be explored, assessing whether the experiential and interdisciplinary approach, even in this case, generates greater interest on this topic. Additionally, it will be interesting to observe how students develop social and collaboration skills and whether the integration of different disciplines promotes cooperation among participants. It will also be important to determine whether the use of software and the need for a solid theoretical foundation genuinely represent barriers for students and how these difficulties can be overcome or mitigated.

In summary, while promising, the proposed approach for upper secondary school will require a thorough evaluation during its experimentation phase. This phase will allow for optimizing the proposal, identifying practical challenges, and assessing its overall effectiveness.

Acknowledgements The authors would like to thank: • Institute "*S. Aurigemma*", Via Nazionale, 13, 83024, Monteforte Irpino (AV) for making the research activities possible. • The student Noemi Ambrosio for her support during the experimentation.

References

Allen M (2019) Misconceptions in primary science 3e. McGraw-Hill Education (UK)
Andriesse DC (2005) Huygens: the man behind the principle. Cambridge University Press
Baldes RW (1978) Democritus on the Nature and Perception of'Black'and'White. In: Phronesis, pp 87–100
Blizak D, Chafiqi F, Kendil D (2009) Students' misconceptions about light in Algeria. In: 11th Education and training in optics and photonics conference. vol 9666. SPIE, pp 437–444
Bruner J (1966) Toward a theory of instruction. Belknap Press, Harvard University Press
Bybee R et al (2015) The BSCS 5E instructional model. NSTA press Arlington, TX
Chi MTH, Slotta JD, Leeuw ND (1994) From things to processes: a theory of conceptual change for learning science concepts. Learn Inst 4(1):27–43
Crawford BA (2014) From inquiry to scientific practices in the science classroom. In: Handbook of research on science education, vol II. Routledge, pp 529–556
De Luca R (2010) Numerical evaluation of the escape time of a classical point particle from an annular billiard. In: Revista Brasileira de Ensino de Física, vol 32
De Luca R (2023) The path of a light ray in a semicircular cavity. Phy Educ 58(6):065018
De Luca R, Lamberti V (2024) Wolfram Community, STAFF PICKS, 6 May 2024. Accessed on: 19 November 2024. url: https://community.wolfram.com/groups/-/m/t/3171161
Dewey J (2013) How we think: a restatement of the relation of reflective thinking to the educative process. Houghton Mifflin
Donahue WH (200) Kepler's optics: letters to herwart von Hohenburg. Cambridge University Press
Fazio et al C (2021) Strategies for active learning to improve student learning and attitudes towards physics. In: Beata J, Csaba S (ed) Teaching-learning contemporary physics : from research to practice. Springer International Publishing, Cham, pp 213–233. ISBN: 978-3-030-78720-2. doi: https://doi.org/10.1007/978-3-030-78720-2_15
Fuchs HU, Corni F (2023) Primary physical science education: an imaginative approach to encounters with nature. Springer Nature
Furtak EM et al (2012) Experimental and quasi-experimental studies of inquiry-based science teaching: a meta-analysis. Rev Educ Res 82(3):300–329
Galili I, Hazan A (2000) Learners' knowledge in optics: interpretation, structure and analysis. Int J Sci Educ 22(1): 57–88
Garritz Andoni (2013) Teaching the philosophical interpretations of quantum mechanics and quantum chemistry through controversies. Sci Educ 22:1787–1807
Glazebrook RT (1896) James Clerk Maxwell and modern physics, vol 6. Cassell and Cie
Grosseteste R, Panti C (2011) La luce. Greco, arabo, latino. PLUS-Pisa University Press. ISBN: 9788884927934. url: https://books.google.it/books?id=xdOIZwEACAAJ
Henriksen EK et al (2018) What is light? Students' reflections on the wave-particle duality of light and the nature of physics. Sci Educ 27:81–111
Huygens C et al (1900) The wave theory of light: memoirs of Huygens, Young and Fresnel, Vol. 15. American Book Company
Kaku M (2004) Einstein's cosmos: how albert Einstein's vision transformed our understanding of space and time. W. W. Norton and Company, New York
Kemp M (2007) Leonardo on Light. Oxford University Press
Kepler J (1937) Ad Vitellionem paralipomena quibus astronomiae pars optica traditur. latino. apud Claudium Marnium, haeredes Ioannis Aubrii-European Library of Information, and Culture (BEIC)—Milan 1604

Kolb DA (2014) Experiential learning: experience as the source of learning and development. FT press

Maxwell JC (1873) A treatise on electricity and magnetism, vol. 1. Clarendon press

Michelini M (2023) La formazione degli insegnanti: un urgente impegno a cui assolvere. In: Societá italiana degli storici della fisica e dell'astronomia: proceedings of the SISFA 42nd annual conference, Perugia, 26-29 Sept 2022, Department of Physics and Geology, University of Perugia. Pisa University Press, pp 101–108

Morin E (2014) La Tête bien faite. Repenser la réforme, réformer la pensée. L'Histoire immédiate. Editions du Seuil. ISBN: 9782021158168. url: https://books.google.it/books?id=z2DEAgAAQBAJ

National Research Council (2000) Inquiry and the national science education standards: a guide for teaching and learning. National Academies Press, pp 14–25

Newton I (1952) Opticks. Dover Publications, Unabridged republication. New York

Schon DA (2013) Displacement of concepts. Routledge

Stefanidou C, Skordoulis C et al (2014) Subjectivity and objectivity in science: an educational approach. Adv His Stud 3(04):183

Strat TTS, Henriksen EK, Jegstad KM (2024) Inquiry-based science education in science teacher education: a systematic review. Stud Sci Educ 60(2):191–249

Testa I, Psillos D, Molohidis A (2020) How national curricula affect the design and transfer of a teaching-learning sequence between two educational systems: case studies from Greece and Italy. Phy Rev Phy Educ Res 16(2):020146

Tóth K, Michelini M, Bitzenbauer P (2024) Exploring the effect of a phenomenological teaching-learning sequence on lower secondary school students' views of light polarisation. Phy Educ 59(3):035009

Voudoukis N, Kalkanis G (2007) Students understanding and teachers education on wave-particle duality–a proposal for scientific to educational models transformation. In: ESERA 2007 Conference, August 21st-August 25th. Laboratory experience of optical reflection in a circular cavity 13

Watson B (2016) Light: a radiant history from creation to the quantum age. Bloomsbury Publishing USA

Witelo (2025) Perspectiva Vitellionis. Bibliothèque nationale de France, Dèpartement des manuscrits, Paris, Latin 7248, 1300

Studying Thermodynamic Processes with Open-Source Hardware and Software Systems

Giancarlo Artiano, Emilio Balzano, and Francesco Guida

Abstract Research in physics education is always open to seizing the opportunities that new technologies can have on the teaching-learning processes. One fascinating area of research centres around students' capacity to cultivate comprehensive skills across scientific domains. This includes designing homemade experimental equipment and coding scripts to analyse and present data. Open-source hardware (OSH), programmable and automated low-cost electronics, have become widespread in recent years, and they can be used not only as tools to support the measurement processes of physical quantities but also as objects to engineer versatile and unique experimental apparatus. Accompanying this process are Open-Source Software (OSS) simulation platforms and programming languages that allow the writing of codes for programming boards increasingly intuitively. The integration of OSH and OSS systems is spreading significantly in institutions designated for training at all levels of education (Heradio et al. 2018). Over the years, our research group has developed experiences in studying the modelling of physical phenomena through the OSH and OSS apparatus. We will present and discuss the study of thermodynamic transformations, construct a gas thermometer and create thermodynamic cycles using Open Source Hardware (OSH) and Open Source Software (OSS) technologies. In conclusion, we will examine the opportunities that the proposal offers for teaching and learning critical thermodynamic concepts such as ideal gases, cyclic transformations, equilibrium states, and quasi-static and reversible transformations. The present study is intended for teachers and students in the final years of high school and those in university laboratory courses.

G. Artiano (✉) · E. Balzano · F. Guida
Università degli Studi di Napoli Federico II, Napoli, Italy
e-mail: giancarlo.artiano@unicampania.it

E. Balzano
e-mail: emilio.balzano@unina.it

F. Guida
e-mail: francesco.guida@studenti.unina.it

I. Testa et al. (eds.), *Connecting Physics Education Research and Practice*,
Challenges in Physics Education, https://doi.org/10.1007/978-3-031-86609-8_9

1 Introduction

Over the years, various studies have been carried out in different countries to explore the teaching and learning of thermodynamics across different educational levels (Warren 1972; Goldring and Osborne 1994; Zemansky 1970). In science education research, thermodynamic processes are a complex and intriguing topic that incorporates concepts from physics, chemistry and mathematics. This interdisciplinary nature highlights the importance of understanding these processes, showing how they are connected to various scientific principles and their significance within the broader educational curriculum (Bain et al. 2014; Tarsitani and Vicentini 1996). For example, in some countries, such as Greece, in-depth analyses have been conducted to better understand the issues related to teaching thermodynamics and to develop more effective teaching methods (Kotsis et al. 2023). The literature in physics education extensively documents the challenges students face in understanding the concepts of the discipline. Decades of research have revealed significant learning difficulties among pre-university students in distinguishing between heat, temperature, internal energy and thermal conductivity (Rozier and Viennot 1991; Tiberghien and Delacote 1978; Yang et al. 2009; Smith and Christensen 2015). Nonetheless, the work of Loverude et al. (2002), Hadfield and Wieman (2010) and Meltzer (2004), as well as studies from several countries, shed light on students' understanding of basic thermodynamics in introductory undergraduate physics courses and upper-level undergraduate courses. All studies show that students' understanding is fragile. These challenges persist even after they complete relevant courses (Nottis et al. 2010). According to Stylos et al. (2021) review, these concepts are perceived as abstract, complicated and poorly related to many applications and real-life situations.

Hence, the challenge of enhancing student comprehension in the field of thermodynamics persists. Creating educational resources and activities to stimulate deep comprehension and conceptual understanding regarding thermodynamic processes, state variables for understanding, and the correlation between microscopic causes and macroscopic effects is a complex endeavour (Leinonenet al. 2015). Numerous studies in the realm of physics education have emphasized the educational significance of employing diverse representations to elucidate concepts in physics (Gardner et al. 2013; Van Heuvelen 1991; Van Heuvelen and Zou 2001). The implementation of innovative teaching methodologies and conceptual survey instruments to evaluate students', such as project- and problem-based learning (PBL) and computer simulations appear to be approaches that work in making thermodynamics teaching more engaging and interactive (Bouquet et al. 2017; Tatar and Oktay 2011; Brown and Singh 2021; Brundage et al. 2024; Firetto et al. 2021; Parobek et al. 2023; Löffler et al. 2018). These approaches can help students visualize thermodynamic processes in action and virtually experience the consequences of changes in the parameters of a system. For example, illustrating how thermodynamic principles apply to everyday phenomena, such as the operation of engines, refrigeration or biological processes, is crucial in making these concepts more accessible and relevant to students. Findings from Meltzer (2004) indicate that emphasizing the creation, interpretation, and manipulation of P-V diagrams depicting different thermodynamic processes may yield substantial educational benefits. Specifically, students could improve their understanding through exercises that involve translating between diagrammatic representations and verbal descriptions of specific processes, particularly in cyclic processes.

Digital tools in thermodynamic teaching

This section presents experiences centred on the widespread adoption of open and freely accessible educational resources, data analysis tools and hardware configurations for data acquisition. This vision particularly considers anticipated technological advancements such as open access to knowledge and the transparent sharing of technical resources. In the (Jackson and Laws 2006), the authors emphasize technological devices' advantages to teaching thermodynamics. Using glass syringes, pressure, temperature and position sensors, they illustrate how the student must familiarize first qualitatively and then quantitatively with this branch of physics, which is much more insidious than mechanics. The authors propose verifying the laws on ideal gases, and at the end of the work, they create a thermodynamic cycle using a heat machine. Onorato et al. (2010) discuss about two experiments are proposed to study Boyle's law and the pressure law in a didactic laboratory. The peculiar feature of the experiments is that the value of the pressure and of the volume are obtained respectively by means of a force and a position sensor, thus allowing students to connect, in an experimental context, mechanics variables, such as force and position, to thermodynamics ones, such as pressure and volume. The above mentioned work (Jackson and Laws 2006; Onorato et al. 2010) discuss on the use of microcomputer-based lab (MBL) sensors which give students immediate feedback by presenting data graphically in a manner that can favour the understanding of the relationship between the variables involved. Moreover, students confront their perception of reality with the limits of a quantitative measurement. However, the MBL tools offered by private companies are often black boxes that do not always allow students to see how the individual parts are integrated. Recent research in physics and chemical education on the using of new low-cost open-source equipment (e.g. Arduino microcontrollers and compatible sensors) are discussed in (Bouquet et al. 2017; Haugen and Moore 2025; Organtini 2018; Huang 2015). In these works, students had complete autonomy in developing their experimental configuration and studying their chosen physics topic. Results demonstrate that these experiments foster student participation and creativity, substantially supporting their learning. A unconventional method to investigate thermodynamic processes with open source systems is proposed in (Essaadaoui et al. 2021). Authors propose an non-traditional laboratory experience to measure the mechanical equivalent of heat using the DS18B20 probe as a temperature sensor connected to an Arduino UNO microcontroller to capture temperature versus time data in a calorimeter. Although several institutions, including schools and universities, have integrated OSH and OSS technologies in their laboratory activities (Heradio et al. 2018), a tiny amount of research still promotes and critically discusses activities of such low-cost technology in the teaching and learning of thermodynamics in the final years of high school and university courses.

2 Design of Educational Activities

Pedagogical approach

In this work, we focus on the idea that technological advances do not automatically enhance teaching practices, like those in thermodynamics. Technical resources and technological tools in science education should not be seen as 'black boxes'; instead, they should be understood as transduction mechanisms. Understanding how these digital and programmable mechanisms function is crucial to recognizing their potential drawbacks and limitations. The present study is part of a broader research (Artiano and Balzano 2023b) to develop educational materials that enhance the learning and teaching of a phenomenon through the lens of using new technologies (e.g. OSH and OSS). In a past master's thesis, we discussed using an accelerometer and a gyroscope to analyse the motion of a bicycle wheel in real-time (Artiano 2018). Instead, a significant aspect of the present proposal, which constitutes part of a master's thesis in physics education, is creating unconventional and homemade experimental apparatus from scratch, utilizing programmable microcontrollers and microelectronic mechanical systems (MEMS) pressure and temperature sensors. The field of thermodynamics and its interdisciplinary nature of content necessitates an instruction approach that integrates theoretical concepts, practical experiences, numerical calculations and the technological use of experimental apparatus. We have developed an experimental setup using OSH and OSS systems to study multiple aspects of thermodynamic processes. The several aspects of the activities are not to be understood hierarchically but as crucial points in a cycle of observation and theoretical reflection in line with the do-it-yourself culture (DIY). Following the DIY approach, students can participate in every aspect of the experimental process: assembling and configuring the hardware based on the experiment, writing or modifying code using the open-source software, managing data transfer to the PC, configuring software for real-time data visualization and performing data analysis.

1. Conducting experiments using common objects to explore various phenomena;
2. Selecting and programming sensors and microcontroller boards using open-source hardware;
3. Designing data communication protocols (via serial, Wi-Fi or bluetooth) between the measurement instrument and a computer;
4. Sampling and analysing data using open-source software.

Experimental setup

The present proposal encompasses experimental activities focused on analysing thermodynamic transformations and validating the working mechanisms of a gas thermometer and a cycle thermodynamic. We will show how the versatility of a single-sensor interface, highlighting its adaptability to various experimental measurements, allows the study of the multiple aspects of the same phenomenological context. Due to its reliability in measurements, low cost and good accuracy, this system has the potential to elucidate several critical topics in thermodynamics. These include:

- Observation of real-time pressure, temperature and volume trends, even for non-equilibrium states.
- Verification of complex theoretical models, such as measuring the absolute temperature with a gas thermometer.
- Manipulating a heat engine to comprehend the first law of thermodynamics by creating a thermodynamic cycle in real-time.

The experimental setup consists of:

1. Common objects for experiment preparation;
2. A core of microelectronics and open-source software for data sampling and analysis;
3. Various data communication channels based on the specific experiment.

The educational activities are indicated in Table 1 while the experimental setup for organizing the learning experiences is represented in Fig. 1.

The experience generally starts with detecting temperature and pressure values of the gas (air) inside the container (or syringe in the case of Boyle's law) by the sensor denoted as BMP180. Subsequently, this data is transmitted via the I^2C protocol to the Arduino microcontroller. The Arduino microcontroller, in turn, dispatches the data in CSV format to a computer through the HC05 Bluetooth module or via serial

Table 1 Content of experiments and their experimental setup

Experiments	Homemade materials	Sensors and microelectronics	Data sampling and analysis
Boyle's law, thermodynamic cycle	50 cm^3 piston syringe, slide potentiometer, rubber tube, 3-way tap for venting the system	Arduino microcontroller board, BMP180 pressure and temperature sensor	Computer for data acquisition via USB cable, KST Graphical Software for Data Plotting
Lussac's law	Airtight container	Arduino microcontroller board, BMP180 pressure and temperature sensor	HC05 bluetooth module with battery for power supply, computer for data acquisition via bluetooth, KST graphical software for data plotting

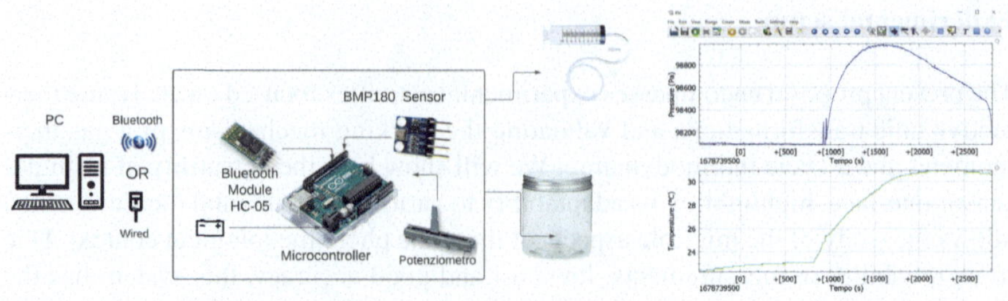

Fig. 1 General scheme of experimental setup and examples of pressure and temperature graphs

communication. The Real-Term serial monitor, installed on the computer, facilitates the management of this transmission protocol. Real-Term is a valuable program that stands out for its versatility, adaptability to various serial communication protocols and ability to present data in different formats. Upon initiating the data sampling process, the data can be observed in real-time using the KST software or a Python script. Notably, the KST software enables the graphical representation of temperature and pressure in real-time as shown in Fig. 1. The real-time data sampling holds significant educational value specially in thermodynamics events, as the graphical representation of data in real-time greatly aids in comprehending the dynamic aspect of the ongoing phenomena. While it is undeniable that the thermodynamic equations referenced are linked to equilibrium states, each state can also be perceived as the inception or consequence of a transformation, and every transformation as a 'path' connecting two states.

3 Educational Activities

In the following sections, we outline four experimental activities conducted with high school educators during a professional development workshop, and later with a group of students enrolled in a physical education course. This documentation has been essential for a comprehensive master's thesis in physical education. The activities described include both the theoretical foundations of the physics experiments and the teaching opportunities they provide for educators and students, thereby enhancing the teaching and learning environment.

Boyle's Law

To analyse the Boyle-Mariotte law, students could design an engaging experimental procedure. This experiment utilized a large, graduated syringe with a capacity of 50 cm^3 to observe the relationship between pressure and volume in a gas. A key setup component was a slide potentiometer integrated at the upper end of the syringe. This

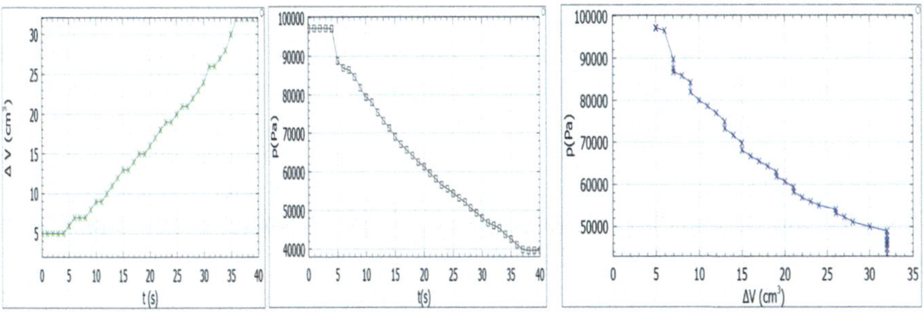

Fig. 2 Real-time plotting of pressure and ΔV as a function of time and Clausius Clapeyron graph

potentiometer is essential for accurately measuring volume variations as it converts resistance values ranging from 2 to 10 kΩ into a 0–5 V signal. A microcontroller then processes this voltage to map the resistance to corresponding volume values, ensuring precise tracking of the movements within the syringe. To effectively measure pressure changes, a rubber hose was connected to the lower end of the syringe, with the other end attached to a pressure and temperature sensor. This sensor provides valuable data on the gas's behaviour under varying conditions. The careful wiring of the potentiometer and sensor to the microcontroller is crucial for accurate data acquisition. The experimental apparatus has been meticulously engineered to enhance the learning experience, aligning with the core principles discussed in the introduction of Boyle's Law. An important assumption made during these measurements is that the number of moles of gas (in this case, air) remains constant throughout the experiment. Before taking measurements, it is essential to allow the system to reach thermal equilibrium with the surrounding environment, which ensures that temperature variations do not affect the results. By manipulating the syringe plunger, students can observe how changes in volume influence the pressure of the gas, providing a hands-on understanding of Boyle's Law. The initial volume $V_0 + 5\,\text{cm}^3$ is increased until it reaches the volume of $V_0 + 30\,\text{cm}^3$. The following Fig. 2 shows the time evolution of the state variables and the Clausius Clapeyron graph.

The estimate of the value of V_0 can be obtained by exploiting the constancy of the product pV, from which we derive the relation:

$$V_0 = \frac{p_2 \Delta V_2 - p_1 \Delta V_1}{p_1 - p_2},$$

where $V_i = V_0 + \Delta V_i$ and the indices *1* and *2* indicate two generic equilibrium states. An estimate[1] for V_0 is $14.5 \pm 2.7\,\text{cm}^3$ and it is compatible with the measured volume V_0 of $14 \pm 1\,\text{cm}^3$.

[1] This estimate was obtained between the points (p_1 = 97142 Pa, $\Delta V_1 = 5\,\text{cm}^3$) and ($p_2$ = 42519 Pa, ΔV_2 = 30 cm^3). The low sensitivity of the potentiometer dictates the high error. We also note that in the continuation of this paragraph, we will use the estimate of V_0 without considering the error ϵ_{V_0}.

From the curvilinear fit, we obtain the value of the constant cost = 1868830 ± 4225 Pa cm^3. While from the measured values of pressure p_i and volume V_i, we obtain a coefficient of statistical variability between the various products $(pV)_i$ of $CV = \frac{\sigma_{pV}}{<pV>} = 0.012$.

From the estimated value of pV = cost, it is possible to estimate the value of the number of moles of gas n_{stim}, and the calculation returns a value of $0.78 \times 10^{-3} \pm 2.5 \times 10^{-5}$ mol. Comparing this value with the value obtained from the equation of state of ideal gases $n_{th} = 0.83 \times 10^{-3} \pm 6.8 \times 10^{-5}$ mol.

This experiment allows students to collaborate with their teachers, reinforcing theoretical concepts while enabling practical application. It provides real-time visualization of changes in various variables, promoting a more comprehensive understanding of gas behaviour and properties. Through this hands-on approach, students are better equipped to understand complex scientific principles and think critically about gas dynamics.

Lussac's law

In this educational experiment, students explore the fundamental concepts of thermodynamics, mainly focusing on the relationship between pressure and temperature in a gas. The proposed methodology illustrates a fully automated measurement process that enhances learning and understanding. To initiate the experiment, students enclose the microelectronic unit—comprising a power supply and sensors—within a vacuum-sealed container. This setup ensures that both the volume V of the gas (in this case, air) and the number of moles n remain constant. By keeping the volume of the container fixed, they can observe the behaviour of the gas during an isochoric process, where volume does not change. The primary objective of this teaching experiment is to validate the linear relationship defined by Gay-Lussac's law, which states that the pressure of a gas is directly proportional to its absolute temperature when volume remains constant and to estimate the value of T_0 using linear regression parameters. To do this, students could establish a controlled environment where the gas can be subjected to varying temperatures. Referring to Eq. 1 we obtain

$$p = \left(\frac{nR}{V}\right)(\theta + T_0) = \text{cost}(\theta + T_0) \tag{1}$$

$$p = \left(\frac{nR}{V}\right)(\theta + T_0) \equiv m\theta + q \tag{2}$$

where

$$m \equiv \frac{nR}{V} \qquad q \equiv \frac{nR}{V} T_0 \tag{3}$$

it is now easy to observe that from q and m we can derive an estimate of: $T_0 = q/m$.

Students vary the temperature by placing the container in a refrigerated environment at temperature T_1 for a predetermined duration. After this period, the container is transferred to a room temperature environment T_2, where $T_1 \prec T_2$. This transition allows the gas within the container to warm up gradually, and we can monitor how the pressure changes in response to increasing temperature. Throughout the process, they measure the temperature-dependent pressure values at regular intervals of 0.5 seconds. These readings are systematically recorded in a file titled *isocora.txt*. By analysing the data collected, they can use linear regression techniques to estimate T_0, the temperature at which the pressure extrapolates to zero.

In about an hour of acquisition, we obtained 7000 measurement points, going from the temperature T_1 = –8.6 °C to T_2 = 16.6 °C and from the pressure p_1 = 88104 Pa to p_1 = 97795 Pa. In the following figures, we observe the real-time graphs obtained with KST for pressure and temperature (Fig. 3).

From the temperature and pressure values shown in Fig. 3 it is interesting to derive the time constants τ_θ and τ_p, through an exponential fit we have $\tau_\theta = 1575.5 \pm 1.9$ s and $\tau_p = 1123.8 \pm 1.4$ s.

From the estimation of the parameters of the linear regression carried out, we obtain the following values: $q = 92790.3 \pm 5.6$ and $m = 310.2 \pm 0.5$ from which we obtain an estimate for the value of $T_0 = q/m = -299 \pm 0.5$ °C.

By limiting the regression to the first 3000 or 3500 measurement points, we obtain respectively $T_0 = -263 \pm 1$ °C and $T_0 = -274 \pm 1$ °C.

They are values consistent with the expected value α = –273 °C. This can be further food for thought to submit to the students: for example, new measurements

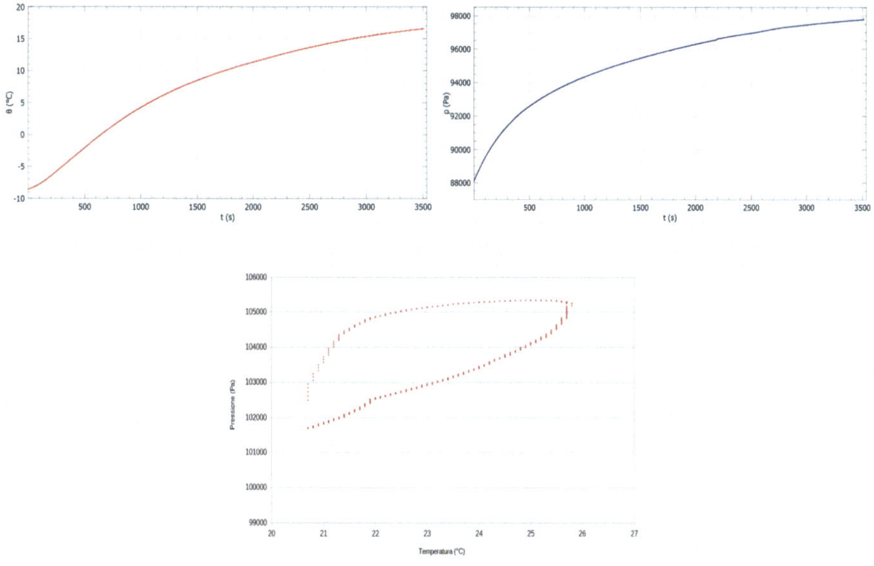

Fig. 3 Graphs of temperature and pressure

Table 2 Temperature estimation using the apparatus as a gas thermometer

p (Pa)	$\theta_{\text{estimated}}$ (°C)	θ_{sensor} (°C)	$\theta_{\text{thermalbath}}$ (°C)
93320 ± 100	9.5 ± 0.4	9.1 ± 0.5	5.8 ± 0.3
93500 ± 100	10.2 ± 0.4	9.8 ± 0.5	6.4 ± 0.3
93650 ± 100	10.7 ± 0.4	10.3 ± 0.5	6.8 ± 0.3

can be made on sub-intervals of temperature, repeat the estimates and compare the results. This experiment solidifies our understanding of gas behaviour under constant volume and provides a practical application of linear regression in deriving critical thermodynamic parameters. By engaging with this experiment, students can develop a deeper appreciation for the principles governing gases and enhance their data analysis skills.

Gas thermometer

This paragraph illustrates the construction and operation of the gas thermometer. Compared to a theoretical model, the system proposed here directly measures pressure through the sensor inside the gas. Similarly to what was done above, we obtain an estimate of the regression parameters q and m that appear in Eq. (1) for a new set of measurements, obtaining m = 302.2 ± 0.4 and q = 90416 ± 6. The straight line thus obtained represents the curve for the estimate of temperatures outside the calibration interval,[2] which in the case in question varies between θ_1 = 12.6 °C and θ_2 =14.4 °C. From Eq. (2), it is easy to deduce that

$$\theta = \frac{p - q}{m}$$

Therefore, it is possible to determine the corresponding temperature at a fixed pressure and vice versa. For this purpose, we have put our apparatus in thermal contact with a thermal bath at different temperatures.[3] After waiting a certain amount of time, for the thermal equilibrium between the bath and the gas contained in the system to be established, the pressure value p returned by the sensor is read. From the parameters q and m of the regression line, we can calculate the corresponding temperature value $\theta(p)$ and compare it with the actual temperature of the thermal bath. Table 2 contains the results for four different measuring points (Fig. 4).

In our observations, we notice that the estimated temperature, denoted by $\theta_{\text{estimated}}$, closely matches the measurement of the sensor value, denoted by θ_{sensor}, within the margins of error. However, the temperature of the thermal bath, represented by

[2] We note that strictly speaking, only two points would be enough to determine the calibration line of a gas thermometer; this is how it is illustrated in physics manuals.

[3] Measured with the HI 98501 thermometer produced by Hanna instruments, this has an accuracy of ±0.3 °C and a resolution of 0.1 °C in the range –30:120 °C.

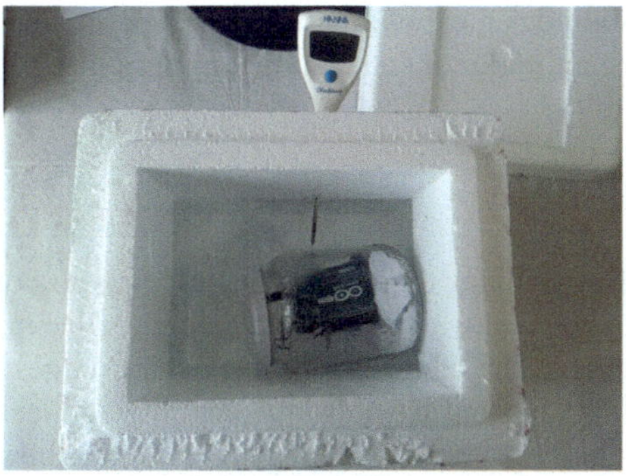

Fig. 4 Experimental setup to test the system functioning as a constant volume gas thermometer

$\theta_{thermalbath}$, is consistently overestimated. One possible explanation is the heat generated inside the gas by the electronics. It is worth noting that the thermometer used to measure the thermal bath temperature and the sensor provide similar temperature readings when they are placed in the same environment.

Thermodynamic Cycle

It is important to note to students how tricky it can be to 'guide' a cycle and thermodynamic transformations in general. Observing the variables pressure (p) and volume (V) in real-time shows that completing a cycle is not a simple task (Fig. 5).

Using the setup described in the experiment referencing Sect. 3, a 50 cm^3 syringe was attached to the container capsule. Initially, the system underwent expansion from V_0 to $V_0 + 50$ cm^3, followed by compression to return the system to the volume V_0.

As can be seen, the cycle does not close perfectly due to an increase in pressure (with the same volume). This presents a valuable learning opportunity. A discussion with the students on the physical reasons for this phenomenon and on where one

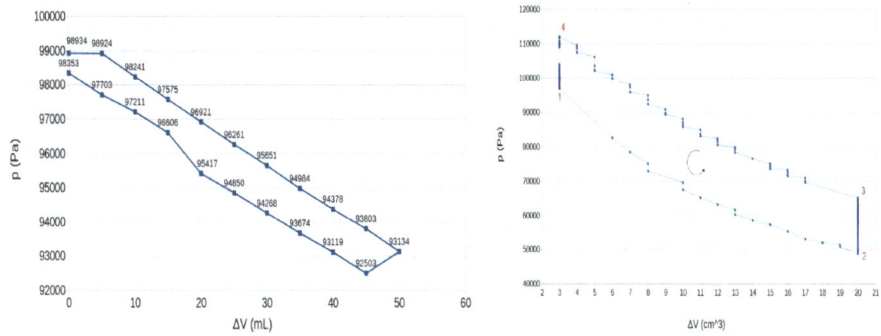

Fig. 5 Thermodynamic cycle real-time

should act, experimentally, to bring the final state close to the initial one turns out to be an important educational opportunity.

Discussion

The primary focus of this study was to construct a distinctive experimental setup using readily available and cost-effective materials to analyse the behaviour of ideal gases. The educational endeavours suggested encompassed the validation of Gay-Lussac's and Boyle's laws and the application of gas thermometry and thermodynamic cycles. The findings corroborate the system's efficacy, as it enables the exploration of various phenomena while delivering results consistent with those obtained from experiments utilising specialized sensors. This proposition is tailored for laboratory students enrolled in master's degree programs, first-year students in scientific disciplines, educators seeking to demonstrate practical experiments in laboratory and classroom settings, and professionals engaged in teacher training.

Some thermodynamic concepts, such as state, internal energy, work and transformations, can be meaningless. For example, pressure can be effectively introduced through practical experiments using glass syringes (Jackson and Laws 2006). One notable feature of Jackson and Laws work is to propose thermodynamic cycle that allows students to experiment with a functioning heat engine. This practical experience makes the abstract concept of a heat engine more tangible, as students physically drive the engine through its cycle while maintaining a focus on quantitative aspects. Several authors in the literature have suggested exploring thermodynamics, particularly in validating gas laws. For instance, there has been a proposal to utilize force sensors as a replacement for traditional pressure sensors (Onorato et al. 2010). This approach was adopted to account for friction between the piston and the body of the syringes during measurements. Furthermore, leveraging modern technological devices to enhance this experience is considered a beneficial element that fosters learning. In line with the pedagogical approach of the research group (Guidoni et al. 2005; Artiano and Balzano 2023a), an experimental apparatus was designed not only to verify the ideal gas laws, but also to serve as an open and adaptable system. This design aims to be flexible and customisable based on the specific needs of the students, with the customization process itself serving as a valuable teaching activity that actively involves students in the design and comprehension of the scientific principles underpinning their research. Our proposal is particularly well-suited for the curricula of fifth-year classes in scientific high schools, technical institutes focusing on science, laboratory courses in bachelor scientific disciplines (e.g. engineering, mathematics, physics) and professional development programs for secondary schools. This work distinguishes itself by emphasizing hands-on experimentation in a laboratory setting. The proposed activities entail the selection of low-cost electronics, the design of experimental setups, the assembly of electronic components and the development of software for data collection and analysis. The primary objective is to facilitate students' comprehensive understanding of the interplay of various components, enhance their manual skills in electronic assembly (e.g. soldering) and foster a solid grasp of fundamental computer science principles for data collection. The proposed apparatus's low cost, compared to those available from science education companies, provides an opportunity for schools with limited budgets, as well as students and teachers who wish to conduct educational experiments independently, thereby promoting more inclusive and equitable physics learning. In this context, the teacher plays a crucial role, employing varied, flexible teaching strategies tailored to students' needs while not neglecting theoretical aspects and building on students' knowledge from previous contexts. It is important to note that the hands-on approach does not diminish the importance of theoretical aspects. Theory and experimentation go hand in hand. Students will choose the variables to observe and have the opportunity to program devices to measure variables such as pressure, temperature and volume in real-time, which will help exercise formal and practical thinking. Students could ask questions such as What is the functional relationship between the variables? During a transformation, how can they distinguish an equilibrium state from a non-equilibrium one?

Possible limitations and how to overcome them

As noted in introduction Sect. 1, there is still limited research into the impact of low-cost technology and OSH and OSS systems on the teaching and learning of thermodynamics. This may be due to an increased cognitive load for students and teachers unfamiliar with programming and electronics. Sensor and microcontroller setup and calibration require technical expertise, which could be challenging in typical high school settings with little or no laboratory space. We are not blinded and know that the present proposal does not represent the solution to the teaching and learning difficulties. We believe that introducing these systems in everyday teaching lessons may have some pedagogical values with more weight than the above limitations. The instruments that compose the experimental setup of the proposal are open systems. Therefore, the extent to which the system must be assembled will depend on the educational objectives. As for the difficulties related to electronics, from the educational point of view, the experimental setup could be presented to the students already pre-assembled. The same applies to the software side, where the codes could be more or less pre-compiled. How much teaching material needs to be pre-assembled depends both on the educational objectives and on the students' level of knowledge of physics, mathematics and technology.

Conclusion

An aspect that we may not have previously taken into account is the potential role of user-controlled devices in influencing reflections on models and theories. Thermodynamics is a phenomenological science in which the time variable t does not explicitly appear in its laws, and there is no quantitative indication of the proximity of non-equilibrium states to equilibrium. Traditional thermodynamics does not represent non-equilibrium states, but with our device, sensors respond over time. Despite the issue of response times, we observe trends that tend towards equilibrium. How should we interpret these results? Reversible transformations present paradoxes and solutions that lead to intriguing reflections (Valente 2019). Students will be faced with the study of real thermodynamic processes by studying the time dependence of state variables and controlling pressure, volume and temperature acquisition times. This experience will allow them to understand better the concept of quasi-static transformation (essential for reversibility), define the model's limits and recognize real-world aspects such as friction and the limitations of the measuring instruments used. We have experienced this unexpected opportunity in seminars with different users (university students, secondary school teachers and colleagues). In conclusion, this teaching proposal serves as an example of how technology can effectively enhance physics learning and foster the development of cross-cutting skills in other scientific disciplines. However, teachers must guide students, encourage critical thinking and integrate technology into the educational context to maximize the pedagogical benefits. The available references and insights are documented in degree and doctoral theses (Artiano 2018), and the codes and apparatuses we design are documented on Creative Commons sharing platforms (Guida 2025).

References

Artiano G (2018) Phenomenological approach to modeling rigid body rotations using exhibits, simulations and MEMS sensors. In: Physics education master thesis work. Università degli Studi di Napoli Federico II

Artiano G, Balzano E (2023a) Teaching and learning physics in formal and informal contexts. Il Nuovo Cimento 100(182):46

Artiano G, Balzano E (2023b) Macchine e meccanismi in fisica e matematica. In: Proceedings of the 43rd SISFA annual conference, Italy, Padova

Bain K, Moon A, Mack MR, Towns MH (2014) A review of research on the teaching and learning of thermodynamics at the university level. Chem Educ Res Pract 15(3):320–335 (Royal Society of Chemistry)

Bouquet F, Bobroff J, Fuchs-Gallezot M, Maurines L (2017) Project-based physics labs using low-cost open-source hardware. Am J Phys 85(3):216–222

Brown B, Singh C (2021) Development and validation of a conceptual survey instrument to evaluate students' understanding of thermodynamics. Phys Rev Phys Educ Res 17(1):010–104 (American Physical Society)

Brundage MJ, Meltzer DE, Singh C (2021) Investigating introductory and advanced students' difficulties with entropy and the second law of thermodynamics using a validated instrument. Phys Rev Phys Educ Res 20(2):020–110 (American Physical Society)

Essaadaoui R, El Moussaouy A, El Hadi M, Ouariach A, Hachmi A, Mommadi O, Bria D (2021) Construction of an educational device for real time data acquisition based on Arduino for a calorimetric study. In: 2021 9th international conference on information and education technology (ICIET), pp 21–25

Firetto CM, Van Meter PN, Kottmeyer AM, Turns SR, Litzinger TA (2021) An extension of the thermodynamics conceptual reasoning inventory (TCRI): measuring undergraduate students' understanding of introductory thermodynamics concepts. Int J Sci Educ 43(15):2555–2576

Gardner M, Greeno JG, Reif F, Schoenfeld AH et al (2013) Toward a scientific practice of science education. Routledge

Goldring H, Osborne J (1994) Students' difficulties with energy and related concepts. Phys Educ 29(1):26 (IOP Publishing)

Guida F. https://github.com/Arturo8978/Tesi_Magistrale

Guidoni P, Iannece D, Tortora R (2005) Forming teachers as resonance mediators. Int Group Psychol Math Educ 3:73–80

Hadfield LC, Wieman CE (2010) Student interpretations of equations related to the first law of thermodynamics. J Chem Educ 87(7):750–755

Haugen AJ, Moore NT. A model for including arduino microcontroller programming in the introductory physics lab. arXiv:1407.7613

Heradio R, Chacon J, Vargas H, Galan D, Saenz J, De La Torre L, Dormido S (2018) Open-source hardware in education: a systematic mapping study. IEEE Access 6:72094–72103

Huang B (2015) Open-source hardware-microcontrollers and physics education-integrating DIY sensors and data acquisition with Arduino. In: 2015 ASEE annual conference & exposition, pp 26–1205

Jackson DP, Laws PW (2006) Syringe thermodynamics: the many uses of a glass syringe. Am J Phys 74(2):94–101

Kotsis KT, Stylos G, Houssou P, Kamaratos M (2023) Students' perceptions of the heat and temperature concepts: a comparative study between primary, secondary, and university levels. Euro J Educ Pedag 4(1):136–144

Leinonen R, Asikainen MA, Hirvonen PE (2015) Grasping the second law of thermodynamics at university: the consistency of macroscopic and microscopic explanations. Phys Rev ST Phys Educ Res 11(2):020–122 (American Physical Society)

Löffler P, Pozas M, Kauertz A (2018) How do students coordinate context-based information and elements of their own knowledge? An analysis of students' context-based problem-solving in thermodynamics. Int J Sci Educ 40(16):1935–1956

Loverude ME, Kautz CH, Heron PR (2002) Student understanding of the first law of thermodynamics: relating work to the adiabatic compression of an ideal gas. Am J Phys 70(2):137–148

Meltzer DE (2004) Investigation of students' reasoning regarding heat, work, and the first law of thermodynamics in an introductory calculus-based general physics course. Am J Phys 72(11):1432–1446

Nottis KE, Prince MJ, Vigeant MA (2010) Building an understanding of heat transfer concepts in undergraduate chemical engineering courses. US-China Education Review, USA. ISSN 1548-6613

Onorato P, Mascheretti P, De Ambrosis A (2010) Mechanical sensors and plastic syringes to verify the gas laws without neglecting friction. Phys Educ

Organtini G (2018) Arduino as a tool for physics experiments. J Phys: Conf Ser 1076(1):012026

Parobek AP, Chaffin PM, Towns MH (2023) Exploring the impact of disciplinary context on students' dynamic transfer of learning when addressing problems that apply the first law of thermodynamics. Int J Sci Educ 45(7):571–592

Rozier S, Viennot L (1991) Students' reasonings in thermodynamics. Int J Sci Educ 13(2):159–170 (Taylor and Francis)

Smith TI, Christensen WM, Mountcastle DB, Thompson JR (2015) Identifying student difficulties with entropy, heat engines, and the Carnot cycle. Phys Rev ST Phys Educ Res 11(2):116–130. American Physical Society

Stylos G, Sargioti A, Mavridis D, Kotsis KT (2021) Validation of the thermal concept evaluation test for greek university students' misconceptions of thermal concepts. Int J Sci Educ 43(2):247–273 (Taylor and Francis)

Tarsitani C, Vicentini M (1996) Scientific mental representations of thermodynamics. Sci Educ 5:51–68

Tatar E, Oktay M (2011) The effectiveness of problem-based learning on teaching the first law of thermodynamics. Res Sci Technol Educ 29(3):315–332

Tiberghien A, Delacote G (1978) Conception de la chaleur chez les enfants de 1 à 12 ans. In: Physics teaching in schools. Taylor and Francis, London, pp 275–282

Valente G (2019) On the paradox of reversible processes in thermodynamics. Synthese 196:1761–1781

Van Heuvelen A (1991) Learning to think like a physicist: a review of research-based instructional strategies. Am J Phys 59(10):891–897 (1991)

Van Heuvelen A, Zou X (2001) Multiple representations of work-energy processes. Am J Phys 69(2):184–194

Warren JW (1972) The teaching of the concept of heat. Phys Educ 7(1):41 (IOP Publishing)

Yang D, Streveler R, Miller R, Roman AS (2009) Repairing misconceptions: A case study with advanced engineering students on their use of schema training modules. In: 2009 annual conference and exposition

Zemansky MW (1970) The use and misuse of the word heat in physics teaching. Phys Teach 8(6):295–300 (AIP Publishing)

History of Physics and Physics Teaching

Can Old-Quantum-Theoretical Description of Physical Reality Be Considered Worth Teaching?

Luisa Lovisetti and Marco Giliberti

Abstract Despite the Old Quantum Theory (OQT) being internationally included in high-school curricula, the vast majority of research asserts that addressing the OQT is not useful for teaching Quantum Mechanics (QM). However, since it is important that physics education has cultural value for personal growth, as well as being integrated into an active learning context, we believe that dealing also with the historical and epistemological reasons behind the genesis of QM would be highly beneficial. This would lead to a deeper understanding of the theory's principles (and, consequently, of its axioms) in addition to aiding better insights into the Nature of Science. This article is thus a reflective work on the feasibility of incorporating aspects of the OQT into the teaching of QM. The concrete possibility and effectiveness of such an approach have already been experimented. Here, we aim to discuss only the general theoretical framework and the motivations within the contexts of educational reconstruction, of cultural understanding of physics, and of active learning that underpin the proposed approach.

1 Introduction

> There is a tendency nowadays for undergraduates to learn their physics completely from textbooks, not becoming acquainted with the original literature and therefore not realizing how the subject grew and developed (ter Haar 1966, backcover).

Starting from the early 2000s, many countries have included Quantum Physics (QP) among the compulsory topics in secondary school curricula (Stadermann 2019, 2021; Krijtenburg-Lewerissa 2017), to acquaint students with modern physics as well as to allow them to discuss aspects of the Nature of Science (NOS) (Stadermann and Goedhart 2020). In the curricula, and also in the textbooks mainly used at a secondary

L. Lovisetti (✉) · M. Giliberti
Department of Physics "Aldo Pontremoli", University of Milan, Milan, Italy
e-mail: luisa.lovisetti@unimi.it

M. Giliberti
e-mail: marco.giliberti@unimi.it

I. Testa et al. (eds.), *Connecting Physics Education Research and Practice*,
Challenges in Physics Education, https://doi.org/10.1007/978-3-031-86609-8_10

school level (or in non-physics majors), QP issues are generally introduced by starting from many models of the so-called Old Quantum Theory (OQT).

The undeniable cultural value of OQT is substantially recognized by everyone (Michelini and Stefanel 2021; Hadzidaki et al. 2000; Weissman et al. 2019). However, many researchers believe that an approach starting from the historical context can hardly be translated into an efficient educational path for high school (Michelini and Stefanel 2021). Indeed, the vast majority of contemporary research in physics education argues that, for effective and meaningful teaching of Quantum Mechanics (QM)[1], especially in STEM degree programs other than physics and mathematics, discussing and addressing the OQT is not a useful choice (Michelini and Stefanel 2021).

The reasons underlying this fact are often motivated by the idea that an educational approach encompassing both the OQT and QM requires excessive time; time that could be more profitably spent focusing solely on QM contents (Michelini and Stefanel 2021). This would allow for a thorough analysis of the theoretical and physical aspects of the theory, often with particular attention to its technological applications. Furthermore, it is often argued that the presentation of the OQT does not directly contribute to the understanding of QM since all the conceptual and formal obstacles associated with quantum aspects should be addressed again by discussing QM (Michelini and Stefanel 2021). Moreover, there is a widespread conviction that introducing semi-classical models of the OQT could even lead to misconceptions among students (Wittmann et al. 2002). Finally, many are of the opinion that transposing both the OQT and QM into educational settings is challenging. This idea is likely fuelled by the fact that, in the vast majority of textbooks (used in high school and in non-physics/mathematics majors), the OQT (and, often, also some aspects of QM, such as Heisenberg's uncertainty principle and the Schrödinger equation) is presented as a narrative on a substantial (and essential) qualitative level of content, employing an approach called "pseudo historical", certainly not conductive to effective teaching from a didactic standpoint (Michelini and Stefanel 2021).

Hence, in the international landscape of research in physics education, the approaches to teaching QP proposed by various research groups, although extremely diverse and varied, tend to share a common aspect: they start directly from QM contents, largely omitting (or significantly reducing) issues related to the OQT, with rare exceptions that focus only on few isolated topics, mainly the black-body spectrum (Navrátil et al. 2013; Kardaras and Kallery 2020; Onorato et al. 2021), the photoelectric effect (Asikainen and Hirvonen 2009; Sokolowski 2013; Niaz et al. 2010; Klassen 2011; McKagan et al. 2009), and Bohr's atomic model (McKagan et al. 2008; Papageorgiou et al. 2016a, b; Derman et al. 2019; Zarkadis et al. 2021; Niaz et al. 2013) (without providing a comprehensive and unified framework).

Therefore, not only at the university level but also for high-school students, many educational proposals are based on the analysis of certain few and carefully selected experiments—such as those on spin, Stern-Gerlach experiments or those concerning

[1] For a general discussion of the educational aspects concerning the teaching of QP, see Michelini and Stefanel (2023).

the behaviour of photon polarization—to highlight anomalies and thus demonstrate the need for a new theory (that is, QM). In general, these proposals require a basic knowledge of classical physics (mainly mechanics and electromagnetism), typical of a science-focused high school, and are aimed at students in the 12th or 13th grade, while the assumed knowledge of chemistry is minimal, essentially limited to the existence of atoms and electrons.

These approaches are logically well-structured but, in our opinion, they deserve further didactic reflections. Indeed, while these experiments are undoubtedly crucial, they cannot be considered as evidence (or cause) of the necessity for a drastic change in worldview, such as that brought about by QM. In other words, they are not sufficient to let us understand why the theory of QM was born and why it is structured precisely in the way it is. They may even risk generating serious confusion and profound misunderstandings regarding the real essence of NOS. In fact, the process of development and shift of paradigm in science does not necessarily stem solely from the difficulty in interpreting isolated "details" or problems (such as in the experiments mentioned earlier). Instead, it is often the result of the continuous accumulation of anomalies, interpretive challenges and *ad hoc* conceptual models over time. It is through this gradual build-up of tension within the pre-existent theory that conceptual revolutions occur, and significant revisions of scientific conceptions are achieved, leading to a change in worldview and the creation of a new reference paradigm (such as the theory of QM). Therefore, while the mentioned experiments undeniably offer valuable insights for reflection and study, it is essential to integrate them with a broader range of examples and concepts in the teaching of QM. Only in this way can we fully grasp the complexity and depth of this theory, and understand its impact on our comprehension of the physical world.

2 Old Quantum Theory for a Cultural Introduction of Quantum Mechanics

From this perspective, we believe that studying also the OQT is beneficial, since it allows to design paths for QP closely intertwined with NOS, creating a significant framework for an approach that exploits active teaching/learning sequences.

Moreover, since education is aimed at being effectively applicable in practice, we believe that starting from the OQT is useful also because the latter is commonly taught in schools today. Teaching QM completely detached from the OQT would indeed take less time than teaching both OQT and QM. However, since the OQT is already presented in school, the time remaining available for QM would be limited in any case. Therefore, from our perspective, it is much better to redesign the OQT and QM together, so that the OQT be functional for QM and allowing sufficient time to cover QM effectively.

For what concerns the risk of misconceptions (Fischler and Lichtfeldt 1992) deriving from OQT semi-classical models (Wittmann et al. 2002) that are still persisting after the presentation of QM, we believe that the issue mainly lies in students not adequately comprehending the theoretical QM aspects and, therefore, subsequently relying on the last "understandable" models they encountered to develop their own reasoning. Indeed, nobody has ever thought to avoid teaching the Ptolemaic system believing this to be an obstacle to understanding that of Copernicus, or to avoid Thomson's atomic model believing it to be an obstacle to grasping that of Bohr. After all, both Copernicus's model and Bohr's model are easily visualizable, and students do not need to rely on previous models in their reasoning. Therefore, it is not manifest that OQT semi-classical models themselves truly lead to misconceptions; rather, in our opinion, these difficulties may indicate a struggle to transition to quantum ideas, suggesting a lack of understanding of quantum aspects themselves rather than confusion caused by classical models. A change of model occurs when a new model is understood (by students) as superior and more powerful than the previous one. Therefore, in our view, the problem is not presenting, for example, Bohr's model (which, in contrast, presents useful advantages (McKagan et al. 2008)), but rather

1. the inadequate discussion of the limitations of such a model (and of OQT models, in general (Savall-Alemany et al. 2019));
2. the lack of a deeper comprehension of the QM structure of the hydrogen atom, which should also lead to a heuristic and qualitative understanding of its physical aspects.

Overcoming semi-classical models is a complex task that can be facilitated by a four-step active learning process.

1. The student engages with a conceptual problem, often derived from the analysis of phenomenology, and attempts to solve it, creating a model, often with the help of historical texts.
2. Then, he/she recognizes that the model is flawed and *ad hoc* for various reasons.
3. Steps 1 and 2 are repeated for the analysis of other "difficult" situations.
4. Finally, a theory is constructed that interprets most of the experimental facts related to the previously constructed/analysed models and other phenomenology, which must also be consistently explained by the theory.

The minimum amount of QM necessary depends on the students' grade level and the depth one aims to achieve. Specifically, once the axiomatic structure of QM is built, we found it important to describe simple systems that are highly significant in physics. For example, in our experimental tests with high-school and university students, we discussed the Mach-Zehnder interferometer in detail, both with and without a phase shifter, two-level systems with and without an interaction Hamiltonian, and three-level systems. We also explored the harmonic oscillator, which we also used to offer a "reasonable" description of the structure of the hydrogen atom.

The approach we propose—aiming not to be strictly historical but rather a physics approach that includes (but that is not limited to) elements of history—should undoubtedly extend beyond just QM, at least from our perspective, and be applied to the entire teaching of physics. Indeed, we believe it is crucial to highlight the historical and conceptual context in which physical theories originated. We think that this approach should be extended to all areas of physics (not only modern ones but also classical ones, such as classical mechanics, thermodynamics and electromagnetism). In fact, it is a way to make physics not only more comprehensible but also more culturally enriching at the same time. Examples of our approach in other contexts, such as mechanics or electromagnetism, with deeper explorations of historical, philosophical and even musical aspects, can be found in Giliberti et al. (2023). This way of proceeding is essential to understand the questions, problems, difficulties and needs from which they have developed. This not only enhances educational effectiveness but also ensures that teaching is not merely a distorted or unfounded reconstruction of scientific history. Often, in fact, the teaching of physics runs the risk of being perceived as detached from reality and limited to abstract and disconnected concepts. Alternatively, as in the case of the "pseudo historical" narrative often proposed for QP, a false reconstruction of a history that never existed or of motivations that were never provided is presented (Kragh 1992; Leone et al. 2018), giving a mistaken view of the NOS and a distorted image of physics, which appears disconnected from reality. Conversely, we believe that an integrated approach that considers the origins of scientific theories and the motivations behind them is essential for a comprehensive understanding of the NOS (Stadermann and Goedhart 2020).

In this article, we will therefore present our considerations regarding some examples of pedagogical reconstruction of QP, which start from the OQT by leveraging it for a deeper understanding of QM. These will be accompanied by motivations aimed at supporting the importance of providing physical content with an appropriate historical framework. From our perspective, this approach is crucial for facilitating a didactic reconstruction of QM that can actively promote the learning of the NOS. This is actually the first part of a two-piece work; however, the article is self-contained, since it explains the reasons and rationale behind the proposal, while details on the experimentations conducted so far and results obtained are discussed in another paper (Lovisetti 2024).

3 Nature of Science and Quantum Physics

As already mentioned, we believe crucial to integrate physics education into a broader cultural framework, emphasizing the NOS. Although past decades of physics education research underscored the importance of integrating NOS into curricula, a notable absence of explicit guidance for incorporating physics history and philosophy into courses (Caramaschi et al. 2022) still remains. In fact, although NOS has long been regarded as a fundamental component of scientific education, empirical studies consistently report a worldwide limited representation of NOS in physics textbooks (Abd-El-Khalick et al. 2017). This deficiency extends also to university curricula, where the dominance of prevailing scientific paradigms often impedes the inclusion of NOS-related aspects, resulting in widespread misconceptions about scientific progress among individuals not pursuing research careers.

Moreover, even today, most school presentations of physics emphasize an alleged "scientific method" based on "rigid" steps which, however, are essentially not implemented in teaching. This method, in its common form, is completely misleading in describing scientific activity (Woodcock 2014). This fact has important implications in teaching; in fact, it contrasts an active teaching/learning method in which the student is supposed to work "like a scientist". Indeed, it is clear that, if the theoretical reference context that should frame scientific activities does not correspond to the development of scientific inquiry, a disconnect in learning is created, causing tensions at a methodological level.

To overcome this situation, the importance of the cultural NOS—that is, of the epistemological aspects of science, of science as a way of knowing, of the cultural values attributed by society to scientific knowledge, and of the values and positioning of man in his relationships with nature and with other human beings—is now recognized as a fundamental element in scientific curricula (Jenkins 2013). The growing demand for a culturally informed physics education, integrating interdisciplinary connections with epistemological and historical aspects, is evident (Galili 2022). Notably, one of the thematic groups within GIREP (International Research Group on Physics Teaching) specifically focuses on studying how to enhance cultural aspects in physics education. This recognition is certainly not recent. In fact, already in 1855, George Campbell, 8th Duke of Argyll and President of the British Association for the Advancement of Science, argued that

> What we want in the teaching of the young, is, not so much the mere results, as the *methods*, and, above all, the *history* of science. How, and by what steps it has advanced; with what large admixture of error every new truth has been at first surrounded; by what patient watchings and careful reasonings; [...]; this is what we ought to teach, if we desire to see education well conducted [...]. It is not merely for the sake of investing the abstractions of science with something of a living and human interest, that we should recall and revive these passages in her history: nor is it merely to impress her results better on the memory, as we fill up from biographies and other sources of information, the meagre page of the general historian. It is for something more than this. It is both that they may be more encouraged to observe nature, and that they may better understand how to do so with effect (Campbell 1856, p. lxxxiii).

Our research aims to build an approach to QM theory closely intertwined with NOS so that it can build a framework for an approach to active teaching/learning sequences. Among the various characterizations of the constituent elements of the NOS (Abd-El-Khalick et al. 2017), we are particularly interested in building awareness on the following intertwined aspects.

1. *Empirical aspect*: empirical scientific statements derive from the observation of natural phenomena, but the observations are mediated by the assumptions underlying the functioning of "scientific" devices, and/or interpreted within pre-existing theoretical frameworks.
2. *Inferential aspect*: inferences are statements about phenomena that are not directly accessible to the senses. Many scientific constructs (such as forces) are inferential;

in fact, they can be accessed and/or measured only through their manifestations or effects.

3. *Creative aspect*: science is not just a rational or systematic activity and creativity is needed.
4. *Problematic aspect*: scientific observations acquire meaning in light of questions and problems derived from precise theoretical perspectives.
5. *Provisional aspect*: scientific knowledge is reliable and durable, but never absolute or certain. Scientific notions are subject to change; reinterpreted in the light of new or revised theoretical ideas.
6. *Theoretical aspect*: science is based on theories that are well-established and internally coherent systems of explanations, which explain many apparently unrelated observations in different fields; they generate questions (usually more than those they are able to answer) and problems. They are often based on assumptions or axioms and assume the existence of unobservable entities. It is impossible to test a theory directly. Scientists make testable predictions based on theories and compare them with observations. An agreement between predictions and observations increases confidence in the tested theory. Disagreements and anomalies, however, act in the opposite direction.
7. *Social aspect*: scientific knowledge is discussed among peers. It influences and is influenced by various cultural elements and spheres, including the social factors, worldview, philosophy, religion, political and economic factors, etc.

4 Three Pillars for Three Research Questions

Hence, from our perspective, it is both feasible and valuable to introduce a coherent teaching/learning sequence for QP focusing also on the OQT, to aid students in comprehending the shift from a period of normal science (in Kuhn's words) towards the genesis of a new theory (QM), thereby fostering a deeper understanding of the NOS (Stadermann and Goedhart 2020; Peters-Burton et al. 2019). The NOS encompasses fundamental characteristics, principles and mechanisms that delineate and influence the development of scientific knowledge. Proficiency in understanding the NOS is crucial for both scientists and the general public alike, as it facilitates informed engagement with scientific concepts, encourages the development of critical thinking abilities and fosters a profound recognition of the dynamic and evolving nature of scientific knowledge.

From the perspective of physics education, QP undeniably resides on treacherous and disjointed terrain. Indeed, the physical motivations that led to QM are often presented through a pseudo historical approach, which hinders students' proper understanding of the NOS; the mathematical aspects are frequently perceived by students as so disconnected from physical reality that, at best, they are merely memorized; the peculiarities of quantum phenomena are often accompanied by crude metaphors, leading to misunderstandings (as well as connections to bizarre, non-scientific issues found on the web). In our opinion, this occurs for QP more noticeably than in the case of classical physics.

As we aim to build an educational framework that is stable, we therefore questioned how many and what kind of foundations should be laid at its base. Considering that the ideal number of legs (or pillars) required for maximum stability is equal to the size of the supporting surface plus one, and since (jokingly speaking) we are studying everything on the educational plane (which, being a plane, is a two-dimensional surface), the ideal number of legs needed for our educational building is three. Thus, we limited ourselves to three legs that encompassed: (1) the physics contents stemming from the OQT, (2) the active approach for a teaching/learning sequence and (3) the exploration of the NOS by means of the historical and epistemological aspects.

Therefore, the research questions that guided our study were as follows:

(RQ1) Can we use some aspects of the OQT to foster the construction of QM as a comprehensible theory with its proper axioms? In the affirmative case, which of them are crucial?

(RQ2) How can we develop an active learning path based on the ideas and considerations outlined above?

(RQ3) What is the effectiveness of including some aspects of the OQT, regarding the NOS?

5 The Didactic Framework: Formal Aspects

Our aim is to present an educational proposal, grounded in cultural considerations, that integrate harmoniously with teaching traditions and has been tested on field with students and teachers. To this end, after reviewing teaching models (Michelini and Stefanel 2021) and university-level QM courses (Akarsu 2010), we have selected a quite standard axiomatic presentation of QM. In this way the transition towards more advanced studies should be smooth, thus ensuring coherence and continuity in students' educational pathways.

In our educational framework, the path from OQT to formal aspects of QM is designed in such a way as to make the axioms emerge from the need to provide a solid mathematical foundation to certain important conceptual elements that arise from phenomenology and previously discussed semi-classical modelling; conceptual elements which we refer to as physical "principles".

5.1 The Didactic Framework: Quantum Principles

The conceptual framework we propose for presenting QM can be adapted to various levels of depth and different educational stages—from the final years of secondary school to STEM undergraduate courses—with or without the reading and discussion of historical excerpts from key original papers, and with varying levels of mathematical rigour. However, we firmly believe that to preserve the cultural authenticity of this approach, textbook-like simplifications should be avoided and the formal structure of the theory must be maintained. Our method begins with a careful examination of the history of physics leading up to the quantum revolution; however, this should not be mistaken for a historical approach, as it does not necessarily adhere to the chronological order of events—in fact, since the development of the course follows an inquiry-based approach, topics may be presented in a slightly different order than the basic chronological one, depending on the emerging needs. For example, issues regarding atomic structure may be introduced before or after the photoelectric effect, depending on students' questions. Moreover, in addition to the historical questions that led to the birth of QM, we also include more recent interference experiments (of course, placing them in their proper "temporal" context).

The goal of our approach is to present a culturally coherent explanation of the conceptual and logical framework of QM, based on an interpretive attitude—focusing on phenomenology and models from the OQT, and more modern experiments involving interference of electromagnetic beams and matter wave optics—rather than on a purely descriptive and prescriptive one.

The physical principles we are about to discuss, although fundamental pieces of information derived from experience, do not represent the independent and essential core of QM, which is instead formed by the axioms. Nevertheless, the principles serve as valuable guidelines that, in an active teaching/learning process, lead to the formulation of the axioms. These axioms, in turn, must respect the principles, in the sense that the physics described by the principles have to be more or less immediately derivable from them (Lovisetti et al. 2023, 2024).

The entire introductory part, which sets the foundation for the development of our physical principles, is based on the historical reconstruction of the OQT. The general underlying idea is that, in the second half of the nineteenth century, Maxwell's nascent electromagnetism immediately posed at least two significant interpretive challenges that led the physics community of the time to question the different perspectives emerging from existing physical theories. On one hand, the incompatibility of electromagnetism with Galilean transformations led to the development of special relativity. On the other hand, the complexities encountered in the thermodynamic description of aspects related to electromagnetic phenomena—such as the thermal equilibrium of radiation in a cavity, the emission of radiation from hot gases, and the interaction between radiation and matter—ultimately paved the way for QM.

Our journey towards principles begins with the explicit discussion of several fundamental thematic blocks related to OQT, each accompanied by its main interpretative *ad hoc* models: (1) the problem of the black-body spectrum (including the laws of Wien, Planck and Rayleigh-Jeans); (2) chemistry between the eighteenth and nineteenth centuries, the discovery of the electron and some early atomic models (Kelvin, Thomson, Nagaoka and Rutherford); (3) the photoelectric effect (mainly through the works of Hertz, Righi, Lenard, Einstein and Millikan); (4) Bohr's model of the hydrogen atom and Sommerfeld's refinement, together with Franck-Hertz experiment; (5) the Compton effect (a historical reconstruction of some of these topics can be found in Giliberti and Lovisetti (2024a, b, c, d, e).

The path continues with the presentation and discussion of other two thematic blocks concerning interference experiments. The first block (6) consists of examining the similarities in behaviour between appropriately prepared beams of matter particles, and mechanical and electromagnetic beams observed in interference and diffraction experiments. These are "high-intensity" experiments, in which particle aspects can be neglected, at least initially, while wave aspects are predominant, especially when interactions or self-interactions are negligible. After the discussion of the Davisson-Germer experiment, de Broglie's historical proposal of matter waves follows. We observe, here, that the experienced reader—but not the novice student—might be confused by the liberty with which we treat electromagnetic beams and matter beams as "similar". In fact, at high intensity, electromagnetic beams are characterized by wave-like behaviour in real physical space, whereas matter beams, at high "intensity", are described by a wave in configuration space, with a vastly higher number of dimensions than physical space. How one can, nevertheless, introduce a wave-like description in physical space also for a high-intensity matter beam is described in our work (Giliberti and Lovisetti 2024f), which lays the theoretical groundwork for our introduction at a level beyond purely qualitative terms.

The path then proceeds with the discussion with another block, (7) of Young's double-slit experiment and of more recent single-quantum experiments, such as single photon interference (Jacques et al. 2005), and the (this time purely conceptual) discussion of light interference through two calcite crystals (Ghirardi 2005, pp. 37–39). Einstein's 1909 paper *On the Present Status of the Radiation Problem* (Einstein 1909)—where it is demonstrated that, by describing the statistical fluctuations of electromagnetic radiation within a cavity, both wave-like and particle-like characteristics emerge simultaneously—is then presented and contextualized; since it is discussed after recent interference experiments, this is another example of how, for the sake of educational coherence, we have not always followed the chronological order in presenting the topics.

That path then continues with some phenomenology concerning electron interference using an electron bi-prism (Merli et al. 1974; Tonomura et al. 1989), of interference of macro-molecules with masses beyond 25,000 Da and consisting of up to 2,000 atoms (Arndt et al., 1999; Fein et al., 2019), experiments with Mach-Zehnder type interferometers (with light and neutrons (Rauch 1995) and the Kapitza-Dirac effect (Batelaan 2007).

These thematic blocks are often discussed starting from a careful analysis of phenomenology. In this regard, it has been noted (Michelini and Stefanel 2021) that an in-depth analysis of experimental aspects can sometimes obscure the discussion of the conceptual aspects of the developing theory. This is a risk that, in our view, must be taken. The soundness of the principles and axioms of a theory as abstract as QM should, in our opinion, be based on a thorough analysis (without delving too deeply into technical details) of what is actually observed in significant experiments, rather than on what our preconceptions might lead us to imagine. Clearly, for this to be possible in the classroom, detailed suggestions and procedures derived from educational research must be explicitly shared by working with teachers and discussing with them the rationale behind our approach.

In addition to the analysis of physical aspects, a broader cultural framework is also outlined; for instance, highlighting how, at the turn of the nineteenth and twentieth centuries, the Western world underwent profound transformations in various fields. These transformations included philosophical developments as well as changes in social structures, political dynamics, economic progress and scientific research endeavours, among others.

5.1.1 Principle I: Wave-Particle Duality

The laws of chemistry, such as multiple proportions, constant ratios and equivalents, naturally led to the concept of the atom, which became crucial for understanding chemical reactions where matter interacted with matter. A discussion was also provided on the largely philosophical reasons why physicists were slower to accept the existence of atoms compared to their chemists' colleagues, taking several decades longer. The discovery of the electron by Thomson in 1897 revealed that atoms are not indivisible, but composite structures. Similarly, the interaction between matter and radiation supported the hypothesis of a particle-like nature of radiation. The explanations of the photoelectric effect (Einstein) and the Compton effect highlighted the particle-like properties in the interaction between radiation and matter, and the validity of the conservation of energy and momentum in these microscopic interactions. Furthermore, modern experiments at extremely low intensities showed the granular nature of both electromagnetic and matter beams (Tonomura et al. 1989; Sala et al. 2019), providing evidence for discrete "quanta" of matter and radiation.

It is important to clarify that the above-mentioned findings do not signify a "crisis" in classical physics, nor do they imply that phenomena such as the photoelectric and Compton effects necessarily require quantized electromagnetic radiation for their explanation (Richardson 1913). Instead, these observations reveal the functional role of certain ideas that contributed to the development of a new theory. Once this new theory was established, it became the reference for interpreting reality, with earlier hypotheses serving as tools in building its theoretical framework.

On the one hand, therefore, with the above observations we put in evidence the particle-like behaviour of matter and radiation. On the other hand, interference and diffraction experiments indicate that both matter and radiation beams exhibit also a wave-like behaviour.

The analysis of Einstein's 1909 paper demonstrated that, in describing the statistical fluctuations of electromagnetic radiation in a cavity, both wave-like and particle-like characteristics must be considered simultaneously. All together our considerations lead to the physical principle:

Principle I: Wave-Particle Duality—Particles such as electrons, atoms, molecules, photons, and others, display both wave-like and particle-like properties.

5.1.2 Principle II: Statistical Aspects

The analysis of interference patterns in single-quantum experiments (such as those involving Mach-Zehnder interferometers, birefringent crystals with or without polarizers, Fresnel biprisms and double-slit diffraction) demonstrates how the observed patterns arise from a statistical distribution. These experiments highlight the need for a probabilistic description of individual quanta. We can therefore state a second principle:

> Principle II: Statistical Aspects—The wave-like features of the theory emerge from a statistical description of a large number of "quanta", each of which can be assigned a probability of being detected in a specific region of space.

5.1.3 Principle III: Quantization

The discussion concerning the structure of matter focuses on the difficulties of constructing an atomic model using only classical assumptions (like in Kelvin's vortex atoms, Thomson's atomic model and Nagaoka's Saturnian model). Bohr's atomic model, later refined by Sommerfeld, marked a significant advance. Along with Planck's solution to the black-body radiation problem, it underscored the need for a quantized description. Einstein's heuristic model of the photoelectric effect and the Compton effect further strengthened the case for introducing the concept of quantization. We can, therefore state a third principle:

> Principle III: Quantization—Certain physical properties, such as the energy levels of harmonic oscillators or angular momentum in Bohr's atom, are quantized. This means they can only take on values that are discrete multiples of a fundamental unit, as opposed to continuous values.

5.1.4 Principle IV: Same Formulas

The formulas used to describe physical quantities—such as that used by Planck for the energy of a harmonic oscillator in the black-body spectrum, by Einstein for the photoelectric effect, and by Bohr for his atomic model (including energy, angular momentum, electron velocity, conservation principles, etc.), as well as the formulas used to interpret the Franck-Hertz experiment—are formally the same as those used in classical physics. To these formulas, appropriate, although *ad hoc*, quantization hypotheses are "simply" added.

> Principle IV: Same Formulas—The formulas used for physical quantities with classical analogues are formally identical to those in classical physics.

5.1.5 Principle V: State "Uncertainty"

If a measurement of a physical quantity G on a given system provides the value g, then immediately afterword, a subsequent measurement of the same quantity G will yield the same value g. However, the statistical nature of measurements on quantum systems suggests that determining the state of a physical system requires access to a statistical ensemble consisting of many systems presumed to be in the same state. Therefore, we propose an additional principle:

> Principle V: State "Uncertainty"—As in classical physics, it is still possible to assign physical quantities to a quantum system. However, from a measurement on a single system, one cannot (in general) determine which was the state of the system itself prior to the measurement.

In other words, whereas in classical mechanics a series of measurements of physical quantities allows the determination of the state of the system, QM is different. Although it is possible to prepare a system according to specific procedures, determining the state is generally impossible without prior knowledge of how the system was prepared. In QM, the state itself is not generally observable.

We note that research literature often chooses Heisenberg's uncertainty principle or Bohr's complementarity principle as fundamental physical principles. However, we find that our alternative approach is more intuitive, essentially equivalent to the aforementioned principles, and, at least potentially, leading to fundamental consequences, such as the no-cloning theorem, much more quickly.

5.2 The Didactic Framework: Axioms of Quantum Mechanics

As usual, formal aspects of QM are synthetically presented as axioms; in our approach, they are constructed upon the just stated principles. The axiomatic formulation is gradually refined along the path, according to the needs of an active learning approach: at first it is slightly simplified, as it pertains solely to observables with discrete spectrum (Weinberg 2013, pp. 52–96); the case of observables with continuous spectrum is suggested as a generalization, and is not initially postulated.

In the following, we will adhere to the well-known notation proposed by Dirac (1939).

(A1) The state of a physical system is represented by a unit vector $|\psi\rangle$ that contains all the available information concerning the system. It is an element of a Hilbert space $\mathcal{H}$ which is associated with the system.

(A2) Every observable physical quantity G is represented by a self-adjoint operator $\hat{G}$ on $\mathcal{H}$.

(A3) (Born's rule). Each of eigenvalues g_k of $\hat{G}$ represents a possible result of a measurement of the quantity G. If $|g_k\rangle$ is the (normalized) eigenvector corresponding to the eigenvalue g_k, and the state of our system is given by

$|\psi\rangle$, then the probability that a measurement performed on the system yields the result g_k is given by: $p(g_k) = |\langle g_k|\psi\rangle|^2$.

(A4) Immediately after a measurement of the quantity G yields the result g_k, the state of the physical system, is given by $|g_k\rangle$.

(A5) When it is not measured, the temporal evolution of a physical system, whose state vector is $|\psi\rangle$, is governed by the equation $i\hbar\frac{d|\psi\rangle}{dt} = \hat{H}|\psi\rangle$, where $\hat{H}$ is a suitable self-adjoint operator.

Obviously, one must then consider how the previous axioms lead to a unified interpretation of many aspects previously difficult to explain and also be used to address new problems, but this is not the topic of our paper which instead concerns the opportunity to effectively insert OQT description of physical reality into the didactic framework for QM (Lovisetti et al. 2023, 2024).

6 Our Experimentations

The approach we have partially outlined and discussed in these pages has already been tested in four trials at the high-school level (Lovisetti 2024): two 15-hour didactic laboratories (held in presence) titled "Old (but Gold) Quantum Theory" and entirely focused on the OQT. One was an extra-curricular activity, based on voluntary registration, addressed to 11th–13th grade high-school students; the other was a deepening activity inserted in the curriculum and addressed to an entire class of 12th secondary school students together with their teacher. Another activity was a 30-hour online course (8 hours of which devoted to the OQT) titled "Principles and Equations of Physics III: Quantum Mechanics" and featuring the OQT as an introduction to QM, addressed to 11th–13th grade high-school students with their teachers on voluntary base. No student attended both the 15 and the 30-hour courses. Moreover, a second edition of the 30-hour online course was held only with teachers. These activities were conducted in the academic years 2022/23, 2023/24 and 2024/25, involving a total of 320 high-school students and 241 teachers.

Through an active-learning approach, the laboratory sessions actively engaged participants in a conceptual exploration of models derived from the OQT. The details and the effectiveness of these experimentations have been extensively discussed in other publications (Lovisetti 2024). From a tangible perspective, the learning outcomes were positive in terms of grades obtained and in the improvement of critical thinking. Furthermore, the Knowledge Integration Construct (KIC) (Liu et al. 2008) highlighted that students were capable of establishing connections between different ideas and concepts in a specific context.

In parallel, the approach was also tested, in three refinements, with 59 students of master's degree in physics or mathematics, during the courses "Preparation of Didactic Experiences".

7 Conclusions

In summary, the title of our paper is *Can Old-Quantum-Theoretical Description of Physical Reality be Considered Worth Teaching?*, a question to which we believe we have substantially responded in these pages. However, we feel it is worth being more explicit.

From our perspective, the answer is that it is certainly advantageous to start from the OQT to better understand QM. Indeed, it is through the OQT, discussed in its historical correctness but proposed for today's students, that significant physical principles can be derived, upon which formal axioms of QM can be constructed (see RQ1).

Furthermore, if, as we believe, research in physics education also has a concrete purpose of improving teaching effectiveness, connecting continuously with the current practices of much of international education certainly aids in the potential implementation of proposals in educational practice.

Our experimentations have been a concrete example of constructing a learning path based on the framework outlined in this work (Lovisetti 2024; Lovisetti et al. 2023, 2024) and have yielded highly satisfactory results in terms of students' understanding of the topics covered and their practical applicability in school, with significant appreciation from teachers in this regard (see RQ2).

Regarding the effectiveness of OQT concerning the NOS (RQ3), within our framework, all seven aspects of the NOS mentioned in Sect. 3 have been addressed. An analysis of the results on this point is still ongoing; however, a series of questions (and their corresponding answers) on the meaning and construction of models have been reported in Lovisetti (2024) and show a significant relationship between disciplinary aspects and the NOS in student learning in all seven highlighted points except the fifth, (i.e. Provisional aspect).

While it is a well-known fact that the OQT alone does not allow for a complete and meaningful educational reconstruction of quantum theory, we are convinced that the question of whether a reconstruction of QP that also incorporates elements of the OQT is pedagogically feasible and effective should be reopened. Based on the discussions in this paper, we believe that such a reconstruction is not only feasible and effective but also appropriate.

References

Abd-El-Khalick F et al (2017) A longitudinal analysis of the extent and manner of representations of nature of science in U.S. high school biology and physics textbooks. JRST 54(1):82–120

Akarsu B (2010) Einstein's Redundant Triumph "Quantum Physics": an extensive study of teaching /learning quantum mechanics in college. LAJPE 4(2):273–285

Arndt M et al (1999) Wave-particle duality of C_{60} molecules. Nature 401:680–682

Asikainen MA, Hirvonen PE (2009) A study of pre- and inservice physics teachers' understanding of photoelectric phenomenon as part of the development of a research-based quantum physics course. Am J Phys 77(7):658–666

Batelaan H (2007) Illuminating the Kapitza-Dirac effect with electron matter optics. RMP 79(3):929–941

Campbell G (1856) Address of the Duke of Argyll, F.R.S. Rep Brit Assoc Advancem Sci 25:lxxiv–lxxxvi

Caramaschi M et al (2022) Mapping the nature of science in the Italian physics curriculum: from missing links to opportunities for reform. Int J Sci Educ 44(1):115–135

Derman A, Koçak N, Eilks I (2019) Insights into components of prospective science teachers' mental models and their preferred visual representations of atoms. Educ Sci 9(2):154

Dirac PAM (1939) A new notation for quantum mechanics. Proc Camb Philos Soc 35(3):416–418

Einstein A (1909) Zum gegenwärtigen Stand des Strahlungsproblems. Phys Z 10:185–193

Fein YY et al (2019) Quantum superposition of molecules beyond 25 kDa. Nat Phys 15:1242–1245

Fischler H, Lichtfeldt M (1992) Modern physics and students' conceptions. Sci Educ 14(2):181–190

Galili I (2022) Scientifc knowledge as a culture: the pleasure of understanding. Springer, Cham

Ghirardi GC (2005) Sneaking a look at god's cards—Unraveling the mysteries of quantum mechanics. Princeton University Press, Princeton

Giliberti M, Albanese F, Lovisetti L (2023) Principles and equations of physics: a multidisciplinary laboratory. J Phys Conf Ser 2750:012028

Giliberti M, Lovisetti L (2024a) Imagination and intuition: the origins of the old quantum theory. Old quantum theory and early quantum mechanics: a historical perspective commented for the inquiring reader. Springer, Cham, pp 27–73

Giliberti M, Lovisetti L (2024b) Thomson's and Nagaoka's atomic models. Old quantum theory and early quantum mechanics: a historical perspective commented for the inquiring reader. Springer, Cham, pp 117–158

Giliberti M, Lovisetti L (2024c) The photoelectric effect and the electron charge. Old quantum theory and early quantum mechanics: a historical perspective commented for the inquiring reader. Springer, Cham, pp 159–228

Giliberti M, Lovisetti L (2024d) Bohr's hydrogen atom. Old quantum theory and early quantum mechanics: a historical perspective commented for the inquiring reader. Springer, Cham, pp 269–312

Giliberti M, Lovisetti L (2024e) The Compton Effect. Old quantum theory and early quantum mechanics: a historical perspective commented for the inquiring reader. Springer, Cham, pp 313–334

Giliberti M, Lovisetti L (2024f) The Schrödinger equation for a non-quantized matter field: a theoretical introduction for physics education researchers. J Phys Conf Ser 2727:012012

Hadzidaki P, Kalkanis G, Stavrou D (2000) Quantum mechanics: a systemic component of the modern physics paradigm. Phys Educ 35(6):386–392

Jacques V et al (2005) Single-photon wavefront-splitting interference. An illustration of the light quantum in action. EPJ D 35:561–565

Jenkins EW (2013) The 'nature of science' in the school curriculum: the great survivor. J Curric Stud 45:132–151

Kardaras I, Kallery M (2020) A teaching module for blackbody radiation using the continuous spectra of stars. Phys Educ 55:045010

Klassen S (2011) The photoelectric effect: reconstructing the story for the physics classroom. Sci Educ 20(7):719–731

Kragh H (1992) A sense of history: history of science and the teaching of introductory quantum theory. Sci Educ 1:349–363

Krijtenburg-Lewerissa K et al (2017) Insights into teaching quantum mechanics in secondary and lower undergraduate education. Phys Rev Phys Educ Res 13:010109

Leone M, Robotti N, Verna G (2018) "Rutherford's experiment" on alpha particles scattering: the experiment that never was. Phys Educ 53:035003

Liu OL et al (2008) Assessing knowledge integration in science: construct, measures, and evidence. Educ Assess 13:33–55

Lovisetti L (2024) "Can Old-Quantum-Theoretical description of Physics be rendered coherent?" A pilot experimentation for high school. Il Nuovo Cimento C 47(5):277

Lovisetti L, Organtini G, Giliberti M (2023) Inducing the construction of formal axioms of quantum mechanics and fostering their comprehension by high school students: the effectiveness of a conceptual approach. Il Nuovo Cimento C 46(6):200

Lovisetti L, Melli E, Giliberti M (2024) "The Elegance of Quantum Mechanics": a didactic path for high school. J Phys Conf Ser 2750:012022

McKagan SB et al (2008) Why we should teach the Bohr model and how to teach it effectively. Phys Rev Phys Educ Res 4:010103

McKagan SB et al (2009) A research-based curriculum for teaching the photoelectric effect. Am J Phys 77:87–94

Merli PG, Missiroli G, Pozzi G (1974) Electron interferometry with the Elmiskop 101 electron microscope. J Phys E Sci Instr 7(9):729–732

Michelini M, Stefanel A (2021) Approaches on T/L quantum physics from PER literature. In: Jarosievitz B, Sükösd C (eds) Teaching-learning contemporary physics from research to practice. Springer, Cham, pp 3–17

Michelini M, Stefanel A (2023) Research studies on learning quantum physics. In: Taşar MF, Heron PRL (eds) The international handbook of physics education research: learning physics. AIP Publishing, Melville, New York, pp 8-1–8-34

Navrátil Z, Dosoudilová L, Jurmanová J (2013) Study of Planck's law with a small USB grating spectrometer. Phys Educ 48(3):289–297

Niaz M et al (2013) Do general physics textbooks discuss scientists' ideas about atomic structure? A case in Korea. Phys Educ 48(1):57–64

Niaz M et al (2010) Reconstruction of the history of the photoelectric effect and its implications for general physics textbooks. Sci Educ 94:903–931

Onorato P et al (2021) Quantitative experiments in a distance lab: studying blackbody radiation with a smartphone. Eur J Phys 42(4):045103

Papageorgiou G et al (2016a) Students' representations of the atomic structure—The effect of some individual differences in particular task contexts. Chem Educ Res Pract 17:209–219

Papageorgiou G, Markos A, Zarkadis N (2016b) Understanding the atom and relevant misconceptions: students' profiles in relation to three cognitive variables. Sci Educ Int 27:464–488

Peters-Burton EE, Parrish JC, Mulvey BK (2019) Extending the utility of the views of nature of science assessment through epistemic network analysis. Sci Educ 28(9):1027–1053

Rauch H (1995) Superposition experiments in neutron interferometry. Il Nuovo Cimento B 110(5–6):557–569

Richardson OW (1913) The asymmetric emission of secondary rays. London Edinburgh Philos Mag J Sci 25(145):144–150

Sala S et al (2019) First demonstration of antimatter wave interferometry. Sci Adv 5(5):eaav7610

Savall-Alemany F et al (2019) Problem-based structure for a teaching-learning sequence to overcome students' difficulties when learning about atomic spectra. Phys Rev Phys Educ Res 15:020138

Sokolowski A (2013) Teaching the photoelectric effect inductively. Phys Educ 48:35–41

Stadermann HKE et al (2019) Analysis of secondary school quantum physics curricula of 15 different countries: different perspectives on a challenging topic. Phys Rev Phys Educ Res 15:010130

Stadermann HKE et al (2021) How high schools teach quantum physics—A cross-national analysis of curricula in secondary education. J Phys Conf Ser 1929:012045

Stadermann HKE, Goedhart MJ (2020) Secondary school students' views of nature of science in quantum physics. Int J Sci Educ 42(6):997–1016

ter Haar D (1966) The old quantum theory. Pergamon Press Inc., New York

Tonomura A et al (1989) Demonstration of single-electron buildup of an interference pattern. Am J Phys 57:117–120

Weinberg S (2013) Lectures on quantum mechanics. Cambridge University Press, Cambridge

Weissman EY, Merzel A, Katz A, Galili I (2019) Teaching quantum mechanics in high-school—Discipline-culture approach. J Phys Conf Ser 1287:012003

Wittmann MC, Steinberg RN, Redish EF (2002) Investigating student understanding of quantum physics: spontaneous models of conductivity. Am J Phys 70(3):218–226

Woodcock BA (2014) "The Scientific Method" as myth and ideal. Sci Educ 23:2069–2093

Zarkadis N, Stamovlasis D, Papageorgiou G (2021) Studying the coherence of students portrayed representations of the atomic structure—Connections with conceptions and misconceptions. Sci Educ Int 32(2):164–171

A Historical-Didactical Reconstruction of Planck's Discovery of the Constant h

Matteo Leone, Francesca Monti, and Nadia Robotti

Abstract Notwithstanding the huge historical literature about the discovery of Planck's constant, superficial and often incorrect introductions prevail even in university textbooks. We retrace in a didactical perspective the main steps of Planck's research for a theoretical derivation of Kirchhoff's universal function, given by the black-body spectrum, from first principles as related to the theories of electromagnetism and thermodynamics, considering all his communications and papers from November 1899 to January 1901.

1 Introduction

The discovery of the constant *h* by Max Planck (1858–1957) (Fig. 1) and the birth of quantum physics as related to the theoretical deduction of Kirchhoff's universal function in 1860 (Kirchhoff 1860) have been the subject of a huge number of historical analyses since 1936, with the work of Rosenfeld (1936), up to very recent publications (Klein 1962; Dougal 1976; Duncan and Janssen 2019; Robotti and Badino 2000, 2001; Robotti 2001; Gearhart 2002; Badino 2015). Nonetheless, even

M. Leone
Department of Philosophy and Education Sciences, University of Turin, Turin, Italy

F. Monti (✉)
Department of Computer Science, University of Verona, Verona, Italy
e-mail: francesca.monti@univr.it

N. Robotti
Department of Physics, University of Genova, Genova, Italy

M. Leone · F. Monti · N. Robotti
Centro Ricerche Enrico Fermi, Rome, Italy

M. Leone
INFN Section of Turin, Turin, Italy

N. Robotti
INFN Section of Genova, Genova, Italy

I. Testa et al. (eds.), *Connecting Physics Education Research and Practice*,
Challenges in Physics Education, https://doi.org/10.1007/978-3-031-86609-8_11

in university textbooks, superficial, oversimplified, if not even incorrect, introductions to the black-body spectrum and Max Planck's formula prevail and these most common and standardized reconstructions (Serway and Jewett 2019; Halliday et al 2021; Walker 2021) do not grasp the true conceptual and epistemological relevance of Max Planck's long work in the historical context of the end of 1800s when the theories of electromagnetism and thermodynamics were gaining ground: the experimental measurement of the black-body spectrum is considered as the starting point and without any historical reference to its evolution; the Stefan–Boltzmann law and Wien's displacement law are presented as experimental findings; Wien's theoretical deduction of the energy distribution function is not mentioned; the so-called "Rayleigh–Jeans" law and ultraviolet catastrophe are introduced as if they were the first tentative classical explanations of the black-body spectrum starting from the modes of a stationary wave inside a box whose average energy is given by the equipartition principle; and Planck's formula becomes the answer to the need to overcome the classical "Rayleigh–Jeans" prediction to fit the experimental data. A search with the free generative tool ChatGPT for "ultraviolet catastrophe" and "Planck's law" confirms that these are the standard versions: "Before Planck's explanation, classical physics, particularly the Rayleigh-Jeans law, attempted to describe black-body radiation"; "The Rayleigh–Jeans law predicted that the energy density of black-body radiation would increase with increasing frequency without limit, leading to what was known as the ultraviolet catastrophe. This prediction contradicted experimental observations"; and "Planck's law successfully resolves the ultraviolet catastrophe predicted by classical physics and matches experimental observations across a wide range of frequencies and temperatures".

In more advanced treatments, Planck's formula is demonstrated following the Rayleigh route by replacing the equipartition principle with the Boltzmann distribution of discrete energy levels for the material oscillators (Zettili 2022) where the need for Planck's constant comes out by comparison with experimental data. Some authors (Serway and Jewett 2019) even introduce the Bohr atomic model before Planck's formula.

Our aim is to retrace the most meaningful steps of Planck's route highlighting some crucial aspects that should not be missing and should rather be emphasized in a correct didactical approach to the birth of Quantum Physics and of Planck's constant, with a special focus on how these subjects could be taught at the higher education level. In particular, we will take into account Planck's research steps since his doctoral thesis (Sect. 2) and we will discuss his search for a derivation of Kirchhoff's universal function from first principles since the last one (Planck 1899) of five communications to the German Physical Society, dated from January 1897 to November 1899 (Planck 1897a, b, c, 1898, 1899) in continuity with the two following papers (Planck 1900a, b) of January and March 1900 (Sect. 3) up to the October 1900 communication to the German Physical Society (Planck 1900c) and the following papers (Planck 1900d, 1901) of December 1900 and January 1901 (Sect. 4). In our historical-didactical reconstruction mainly meant for the higher education level, we shall put into evidence the conceptual and mathematical cornerstones of his route toward the discovery of the universal constants h and k from the fundamental laws of electromagnetism and

Fig. 1 Max Planck around 1910 (public domain image)

thermodynamics and the role of the interplay between mathematics and physics and between theory and experiments in the growth of science.

2 The Premises: First Planck's Studies About Irreversibility and the Electrical Harmonic Oscillator

Planck's research activity finds its roots in his doctorate thesis "On the second fundamental theorem of the mechanical theory of heat" (1879) and is since then strictly connected to his belief in the universal validity of the second law of thermodynamics as related to the irreversibility of spontaneous processes and to the concept of entropy (Gearhart 2002).

Two fundamental didactic aspects can already be put into evidence. The first one concerns the fact that thermodynamics should be presented as a universal theory and its universal character should be introduced since its very first treatment. The second one concerns the fundamental and profound connection between quantum physics and thermodynamics that is neglected or even forgotten in the most common approaches to the birth of Quantum Physics. Both these two points can be addressed if the birth of quantum physics is introduced following Planck's route that led him to the discovery of *h*.

The irreversibility of radiative processes, i.e., the investigation of how damped electric dipoles, as harmonic oscillators ("resonators") emitting and absorbing electromagnetic radiation, reach a stationary irreversible state, is the subject (and the title as well) of a series of five communications by Max Planck to the Prussian Academy of Science, published between February 1897 and November 1899 (Planck 1897a, b, c, 1898, 1899). These five papers indeed came after two previous fundamental studies devoted to the physics of the damped and forced electrical harmonic oscillator, following Hertz (Planck 1895, 1896).

It is worth noticing an important point that is usually neglected in textbooks even at the university level: the role that, before the discovery of the electron and before the birth of the first atomic models, the study of the damped and forced harmonic oscillator has in Planck's work, since this is his fundamental road that allowed linking, for each frequency, the spectral energy density of the electromagnetic waves at thermal equilibrium, u_ν, to the average energy of the resonator, $\overline{U}_\nu$:

$$u_\nu = \frac{8\pi \nu^2}{c^3}\overline{U}_\nu \tag{1}$$

Equation (1) was not found by Planck starting from the behavior of the stationary modes of an electromagnetic wave inside a cavity, that was in fact proposed only later (Rayleigh 1900; Jeans 1905), as it is commonly shown in textbooks. Details on Planck' calculation of the fundamental equation in the stationary case that led to (1) can be found in Darrigol (1992).

3 The Black Body as Theoretical and Experimental Reference up to the Beginning of 1900s and Planck's November 1899, January 1900, and March 1900 Papers

The starting point of Planck's research was the existence, established by Kirchhoff (1860) of a *universal* function ε_λ expressing the ratio between the emissive power (or emissivity) at a given wavelength e_λ—i.e., the amount of electromagnetic energy emitted per unit time per unit area in a unit solid angle ($e_\lambda d\lambda = (c/4\pi)u_\lambda d\lambda$—and the absorbing power (or absorptivity) a_λ—i.e., the fraction of the incident electromagnetic energy that is absorbed—at that wavelength

$$\varepsilon_\lambda = \frac{e_\lambda(T)}{a_\lambda(T)} \tag{2}$$

that is the same for all bodies at the same temperature independently from their nature and shape.

Kirchhoff also introduced the concept of the "black body", i.e., a body for which $a_\lambda = 1$ that absorbs all the electromagnetic radiation incident on it. Since for a black body $\varepsilon_\lambda = e_\lambda$, the search for the universal function was reduced to the search for the black-body emissive power, hence the key role of the black-body spectrum.

In 1884, L. Boltzmann theoretically deduced, using thermodynamics and the properties of electromagnetic radiation, that the total emissive power of the black body, i.e., the emissive power integrated at all wavelengths, depends on the fourth power of the absolute temperature (Boltzmann 1884), a formula that was previously found by Stefan (1879) by building on John Tyndall's measurements (1864) of the infrared radiation of a platinum filament:

$$\int_0^\infty e_\lambda d\lambda = \text{const} \cdot T^4 \tag{3}$$

Later, in 1893, Wien investigated how the black-body emissivity spectrum changes as the temperature changes and derived *theoretically* that the same specific feature (e.g., the maximum) of the black-body spectrum, at a certain temperature, shifts with the temperature following a "displacement law" based on a feature-dependent constant, $\lambda T = \text{const}$ (Wien 1893).

Moreover, it is here worth noticing for the next developments that Wien was also able to deduce that the black-body emissivity spectrum as a function of λ, $e_\lambda d\lambda$, should depend on the product (λT) or, equivalently, $e_\nu d\nu$ as a function of frequency ν should depend on (ν/T).

In 1896, Wien derived an expression for the black-body emissivity:

$$e_\lambda d\lambda = \frac{C\lambda^{-5}}{e^{c'/\lambda T}} d\lambda \tag{4}$$

where C and c' are two constants whose value Wien did not specify (Wien 1896).

It should be emphasized that Wien's derived (4), now referred to as Wien's energy distribution law, from purely theoretical considerations, starting from three purely theoretical results (the Stefan–Boltzmann equation, Wien's displacement law, and the dependence of $e_\lambda d\lambda$ on the product (λT)) choosing as the radiating system a gas of molecules for which the kinetic theory of gases and the Maxwell–Boltzmann distribution of velocities hold: the temperature was related to the kinetic energy of the gas molecules and the wavelength of the emitted radiation was assumed to be a function of the velocity.

In 1896 also F. Paschen, based on his own experimental data (range of $T = 390$–1397 K, range of $\lambda = 1$–8 μm), obtained, independently from Wien, an empirical law like Eq. 4, and estimated the constant $c' = 1.4455$ °C cm (Paschen 1896).

A number of empirical laws existed. However, Wien's was the one that best fitted with the experimental data and because of this Planck will attempt a theoretical derivation of such a law. Wien's displacement law and (4) showed indeed a good agreement with the first best measurements of black-body radiation made by O. Lummer and E. Pringsheim and reported by them on February 3, 1899 at a session of the German Physical Society in Berlin (Lummer and Pringsheim 1899a) (range of

$T = 621$–1646 K, range of $\lambda = 1$–6 μm). Later that year, however, at the November 3, 1899 session of the German Physical Society, Lummer was now certain, after having improved their methods that there were "discrepancies of a systematic nature between theory and experiment" (Lummer and Pringsheim 1899b, p. 222) and illustrated this by the curves in Fig. 2. In the meantime, unlike Lummer and Pringsheim, Paschen did not find any deviation from Wien's law, as he communicated in two 1899 sessions of the Prussian Academy (held on April 27 and December 7), and the situation remained at this stage undecided (Paschen 1899a, b). Max Planck entered the subject at this point in November 1899, i.e., *before* the measurements (Lummer and Pringsheim 1900; Rubens and Kurlbaum 1900) that would have shown a systematic divergence from Wien's law: Planck's interest was in the "quest for the absolute" and he started his work from Wien's law since he considered it the best theoretical representation of the black-body radiation. With reference to Kirchhoff law, in his scientific autobiography he reminds that "*According to this law the black body energy distribution is a universal function of T and λ. Then it represents something absolute. And because I always considered the quest for the absolute as the highest of all scientific activities, I set myself actively to work* [authors' translation]" (Planck 1948).

From the historical and didactical perspective, Planck's *research program* was to find how the Kirchhoff *universal* function $e_\lambda d\lambda$ or $e_\nu d\nu$ could be derived from first principles, namely, from the second law of thermodynamics, considering that the Kirchhoff universal function is given by the electromagnetic spectral energy distribution in the stationary state of any closed system where the electromagnetic radiation is interacting with matter.

Planck faced the problem starting from Kirchhoff's "cavity theorem", i.e., from the fact that "the radiation emitted by a black body at a temperature T is the same as the thermal radiation in equilibrium into a cavity, with completely reflecting walls and maintained at the same temperature T, independently of the substance which forms the walls or of the substance eventually enclosed into the cavity [authors' translation]" (Kirchhoff 1860). Planck's starting point was, then, that the stationary state of a closed physical system is in turn uniquely identified by its absolute thermodynamic temperature, defined in classical thermodynamics through the *universal* relationship between its average energy (U) and entropy (S)

$$\frac{dS}{dU} = \frac{1}{T} \tag{5}$$

and assumed that the walls of the cavity are made of Hertz harmonic oscillators (resonators) whose energy density $u_\nu\, d\nu$ is given by (1) and $e_\lambda\, d\lambda$ can be derived from the geometrical factor $c/4\pi$ and using $\lambda\nu = c$:

$$e_\nu d\nu = (c/4\pi)u_\nu d\nu \tag{6a}$$

$$\begin{aligned} e_\lambda d\lambda &= (c/4\pi)u_\lambda d\lambda = (c/4\pi)u_\lambda(d\lambda/d\nu)d\nu \\ &= (c/4\pi)u_\nu\left(c/\nu^2\right)d\nu = \left(\lambda^2/4\pi\right)u_\nu d\nu \end{aligned} \tag{6b}$$

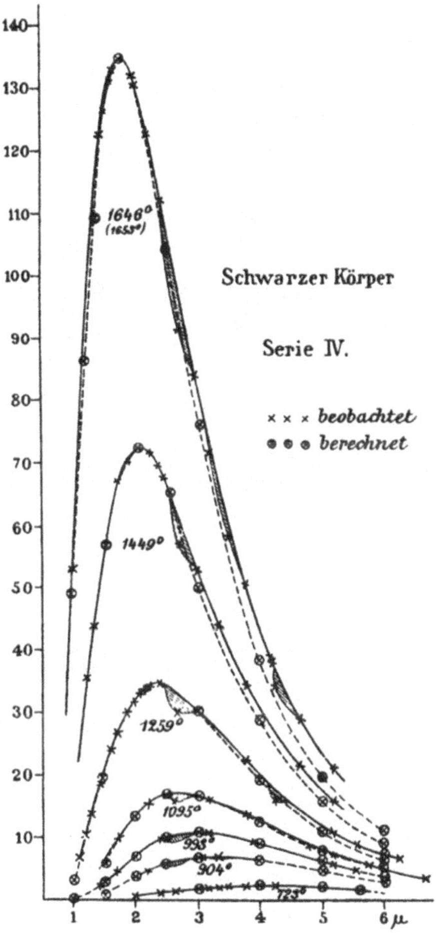

Fig. 2 Lummer and Pringsheim's energy curves (original public domain figure published by Lummer and Pringsheim (1899b, p. 217). The crosses indicate the experimental points while the dots indicate the values predicted by Wien formula. Public domain image

With reference to the didactical analyses proposed by Branchetti et al. (2019) it must be emphasized that Eq. (5) (coming out from thermodynamical principles) and (1) (derived by Planck himself) are two cornerstones of Planck's *mathematization* procedure. A third mathematization cornerstone, namely, the dependence of the entropy from the (U/ν) ratio, will emerge later, in Planck's final derivation of the Kirchhoff universal function (Sect. 4).

In the first phase of his work, Planck assumes that the Kirchhoff universal function is known and expressed by Wien's Eq. (4), so that his *research program* becomes finding by pure theoretical reasoning based on fundamental physical principles the functional dependence of the entropy from the average energy and from the frequency $S\,(U_\nu, \nu)$ that 1) obeys fundamental thermodynamical constraints and 2) leads to (4) by derivation (5), given (1).

Following this unifying approach, Planck's procedure appears to be based on three assumptions:

- the black-body distribution function is a universal law and must contain only "universal constants";
- acceptance of Wien's law implies that these universal constants are two.
- if the black-body distribution has to be derived from the entropy, then these two universal constants must be already present in the expression of the entropy as a function of the energy of the oscillators.

In the already cited communication of November 1899 (Planck 1899) and in the following paper published in January 1900 (Planck 1900a), Planck conjectured a relationship between the resonator entropy and the resonator average energy, based on purely *technical* steps:

$$S_\nu = -\frac{U_\nu}{a\nu}\ln\frac{U_\nu}{eb\nu} \tag{7}$$

where e is the base of the natural logarithms, and a and b are the two universal constants. By taking the derivative (7) with respect to U and putting the result equal to $1/T$, Planck obtained

$$U_\nu = b\nu e^{-a\nu/T} \tag{8}$$

that leads from (1) to the desired distribution law for the energy density (9):

$$u_\nu d\nu = \frac{8\pi\nu^3}{c^3} b e^{-a\nu/T} d\nu \tag{9}$$

Equation (9) (energy density, as a function of the frequency) is in fact equivalent to (4) (emissivity, i.e., energy emitted in the unit of surface in the unit of time and in the unit of solid angle, as a function of the wavelength) that can be obtained multiplying (9) by the geometrical factor $c/4\pi$.

$$e_\lambda d\lambda = \frac{2c^2 b}{\lambda^5} e^{-ac/\lambda T} d\lambda \tag{10}$$

So, with (7) Planck's objective of finding $S(U)$ was accomplished. The new constants a and b replace Wien's constants with $C = 2c^2 b$ and $c' = ac$. As thoroughly discussed elsewhere (Darrigol 1992; Dougal 1976; Duncan and Janssen 2019), Planck was able to give the values for his two "universal constants" a and b:

$$a = 0.4818 \times 10^{-10}\,\text{s}\,^\circ\text{C}, \qquad b = 6.885 \times 10^{-27}\,\text{erg s}$$

Here the "universal constant" b in 1899 had approximately the same value as the more famous "universal constant" h as it was later calculated by Planck in 1900 (see Sect. 4). Planck concluded his November 1899 and January 1900 papers with a paragraph where he proposed for the first time a new system of "*natural units of measurements*" for length, mass, time, and temperature that, in contrast with "*all the systems of measurements adopted so far* [, *was*] *based on a coincidence of random circumstances*". According to Planck, constants *a* and *b* were to be considered "universal constants", on a par with the two universally recognized constants *c* (the speed of light) and *G* (the gravitational constant). As it was written by Planck: "*It cannot be without interest to note that constants a and b (…) provide the possibility of identifying the new units of length, mass, time, and temperature, which, independently of specific bodies or particular circumstances, necessarily maintain their meaning at all times and for all cultures, even for extra-terrestrial or non-human life forms, and can therefore be considered as "natural units of measurements* [authors' translation]" (Planck 1900a).

At the higher education level, Planck's deduction of (9), in the two November 1899 and January 1900 papers, deserves attention as a meaningful example of the role of the interplay of Mathematics (in terms of both mathematization and technical aspects) and Physics (in terms of fundamental principles and physical interpretation) in the growth of science, a point that was addressed also by Branchetti et al. (2019) in the context of the discovery of the Planck's constant, although with a specific focus on Planck's papers after March 1900 and without any reference to his previous work. More precisely, from Eq. 1 and from Wien's law expressed in terms of the frequency (Eq. 9), it follows that U_ν must be given by Eq. (8). Obtaining Eq. (8) from the physical requirement expressed by Eq. (5) requires in turn finding which function of the frequency is equal to $1/T$. The only way to do this is to calculate the logarithm of Eq. 8, leading to the equation $1/T = -(1/a\nu)\ \ln\ (U_\nu/b\nu)$. At this point, it is recognized that $dS_\nu/U\nu = 1/a\nu\ \ln\ (U_\nu/b\nu)$ and S_ν can be obtained through integration as $S_\nu = -(U_\nu/a\nu)\ \ln\ (U_\nu/b\nu) + U_\nu/a\nu$, i.e., Eq. (7).

Yet, (7) was a conjecture, and a physical explanation was still missing. By the way, from (7) the second derivative of the entropy can be obtained

$$\frac{d^2S}{dU^2} = \frac{\text{const}}{U} \tag{11}$$

with the constant given by $(-1/a\nu)$.

In his March 1900 paper, Planck made a strong effort to demonstrate that (11) represented the theoretical and fundamental basis to his conjecture (7), showing that the second law of thermodynamics leads to the constraint that the second derivative of the resonator entropy with respect to its average energy cannot be positive and adding some reasonable requirements about the resonators (Planck 1900b).

4 The Black Body as Theoretical and Experimental Reference Between the Beginning and the End of 1900 and Planck's October 1900, December 1900, and January 1901 Papers

A few months after Planck's derivation of Wien's formula, O. Lummer and E. Pringsheim in early 1900 (Lummer and Pringsheim 1900) (range of $T = 287$–1772 K, range of $\lambda = 12$–18 μm) (Fig. 3) and H. Rubens and F. Kurlbaum in October 1900 (Rubens and Kurlbaum 1900) (range of $T = 85$–1773 K, values of $\lambda = 24$, 31.6 and 51.2 μm) obtained new experimental data, showing that the divergence from Wien's law previously found by Lummer and Pringsheim (1899a, b) was indeed systematic in the long wavelength range.

Once recognized that the Kirchhoff universal function at long wavelengths was not given by Wien's energy distribution law (4), Planck's research program became to directly find from first principles the Kirchhoff universal function that should be expressed by Wien's energy distribution law at shorter wavelengths.

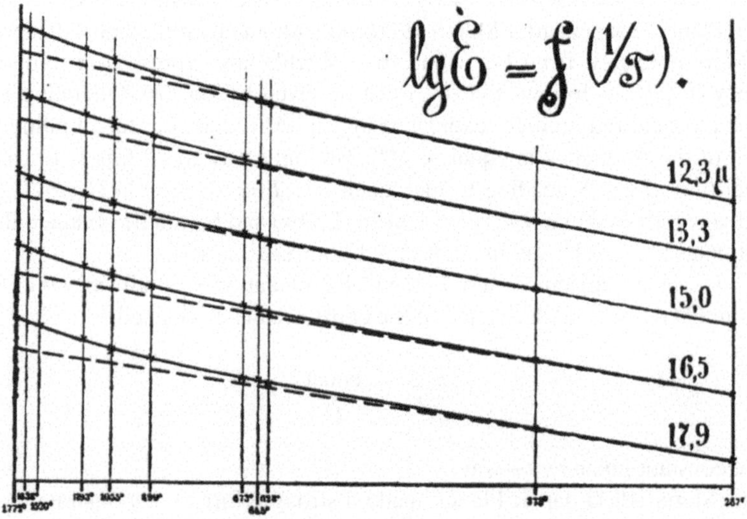

Fig. 3 Lummer and Pringsheim's new measurements of the emitted energy (logarithmic scale) as a function of ($1/T$) in the long wavelength range 12.3–17.9 μm showing a systematic divergence from Wien's law (original public domain figure published by Lummer and Pringsheim (1900, p. 170). The range of values for the temperature on the $1/T$ horizontal axis is, from right to left, 287–1772 K. As it was later observed by Rubens and Kurlbaum, "the contrast between theory and experiment is especially marked when one considers the so-called isochromatic curves, which express the intensity of radiation as a function of temperature for any given wavelength" (Rubens and Kurlbaum 1900, 335–336). As the diagram shows, for sufficiently high values of the product λT, the deviation from Wien's formula is very considerable

In October 1900, Planck presented to the German Physical Society a brief, extremely synthetic paper, in which he proposed a new formula for the energy distribution law (Planck 1900c). Since the new measurements of black-body radiation showed that for low values of the frequency the emitted intensity is proportional to the temperature, Planck deduced that at low frequencies the average energy of the resonator should be proportional to the temperature, hence the first derivative of the entropy in (5) should be inversely proportional to the average energy, leading to $d^2S/dU^2 \sim 1/U^2$ So, he first modified (11) to get the correct dependence both at low and at high frequencies:

$$\frac{d^2S}{dU^2} = \frac{\alpha}{U(\beta + U)} \tag{12}$$

where α and β, again, are two universal constants.

Planck wrote: "This was the simplest of all expressions that give the expected logarithmic dependence from U as expected from probabilistic considerations and reduces to the Wien's law at low wavelengths" (Planck 1900c, 1972, p. 36). It is of great interest to note that Planck adopts here, for the first time, probabilistic considerations to suggest the direction for the physical interpretation of his new conjecture, thus anticipating his famous following papers of December 1900 and January 1901.

Then, Planck introduced his new formula for the distribution law, differing from Wien's 1896 one by a term "−1" in the denominator:

$$e_\lambda = \frac{C\lambda^{-5}}{e^{c'/\lambda T} - 1} \tag{13}$$

where C and c' are the same constants as in Wien's law both in name and in value.

Although Planck did not show the detailed mathematical steps leading from (12) to (13), it is useful to make these steps explicit. It can be easily found by integration of (12) that

$$\frac{dS_\nu}{dU_\nu} = \frac{\alpha}{\beta}\ln\left(1 + \frac{\beta}{U_\nu}\right) \tag{14}$$

and equating this to $1/T$ and inverting the equation gives

$$U_\nu = \frac{\beta}{e^{\beta/\alpha T} - 1} \tag{15}$$

hence

$$u_\nu = \frac{8\pi\nu^2}{c^3}\beta\frac{1}{e^{\beta/\alpha T} - 1} \tag{16}$$

At this point, from Wien's requirement (Wien 1893) already mentioned in Sect. 3 that $e_\nu d\nu$ should depend on ν/T, it immediately follows that β must be proportional to ν ($\beta = B\nu$) and (13) is obtained converting u_ν into e_λ:

$$e_\lambda = \frac{2c^2 B}{\lambda^5} \frac{1}{e^{Bc/\alpha\lambda T} - 1} \tag{17}$$

We can suppose that it was based on these mathematical steps that Planck found the new theoretical expression for the Kirchhoff law (15) presented in the October 1900 communication, valid also at long wavelengths, that substitutes (4). Nonetheless, Planck did not claim he used Wien's already deduced (ν/T) dependence of the *energy distribution* to justify (13). Interestingly, he refers instead to the dependence from (U/ν) of the *entropy* of the resonators, here cited, without demonstration, for the first time, thus again anticipating his following papers of December 1900 and January 1901.

Of course, there is a precise relationship between the two constants α and β (or α and B) and Wien's constants $C = 2c^2 B$ and $c' = \mathrm{Bc}/\alpha$, although we don't know whether Planck performed these calculations. It is important to stress that the two constants are surely *not* the result of interpolating experimental data.

Yet, again, (12) (as it was (7) in the previous derivation in 1899) was a conjecture. By the way, it is of didactical interest to show that integrating (12) and (14) leads to an expression for the entropy (that was not explicitly shown in the October 1900 communication) as a function of the average energy of an oscillator that coincides with the one that will be finally derived by Planck from fundamental physical principles in December 1900:

$$S = \alpha\left[\left(1 + \frac{U}{\beta}\right)\log\left(1 + \frac{U}{\beta}\right) - \frac{U}{\beta}\log\frac{U}{\beta}\right] \tag{18}$$

So, in October 1900, Planck still needed to give a physical meaning to the new law and to derive (13) from first principles. As he himself stated in his Nobel lecture: "*I busied myself, from then on, that is from the day of its establishment with the task of elucidating a true physical character for the formula (...) until after some weeks of the most strenuous work of my life light came into the darkness and a new, unexpected perspective opened up before me*" (Planck 1920).

Following this direction, in December 1900 (Planck 1900d) and January 1901 (Planck 1901) papers, Eq. (18) assumes the role of the starting fundamental physical relationship, in place of (12) for the second derivative, improving previous (7). Planck directly gives its derivation from the fundamental definition of entropy as a logarithmic function of the number of "complexions", that is the number of possible permutations of a finite number of energy elements over the material oscillators and using combinatory calculus, following the "Boltzmann's trend of ideas". Boltzmann (1877) was, in fact, the first one to propose a probabilistic definition for entropy, doing calculations for the specific case of a monoatomic gas.

To this respect, it is didactically relevant to quote Planck's justification for his calculation (Planck 1901, p. 555), i.e., the assumption that the time average of the energy of a single resonator at a certain frequency (which is the meaning of U_ν in (1)) must be equal to the average of the energies at any given time over all the resonators at that frequency: "*The constant energy U of a single stationary oscillating resonator is to be understood as a temporal average or, which results to be exactly the same thing, as the simultaneous average of the energies of a large number N of identical resonators that are situated in the same stationary radiation field, sufficiently far apart so as not to directly affect each other*" [authors' translation].

Planck considered a system composed of N identical resonators at the same frequency ν with an average energy U and total energy $U_N = NU$, an average entropy S and total entropy $S_N = NS$ and generalized the Boltzmann expression using the known absolute constant of the ideal gas, R, and the Avogadro number N, leading to a new universal constant, $k = R/N$, not a priori linked to the gas properties:

$$S_N = k \ln W + \text{const.} \tag{19}$$

In this way, the expression for the entropy, usually known as Boltzmann formula, was written by Planck for the first time, and represents the fundamental and universal physical relationship at the basis of Planck's final theoretical derivation of the black-body emissivity spectrum.

The next step was to calculate the number of complexions W, which requires counting, and counting requires to introduce finite elements of energy ε. Planck assumed that $U = P\varepsilon$ so that also $P/N = U/\varepsilon$) and gave an example of one of the possible complexions (Fig. 4) in the case of $N = 10$ resonators (labeled in the upper line of the figure) and $P = 100$ energy elements (distributed among the resonator as shown in the lower line of the figure):

In this way, the number of complexions, i.e., the number of possible ways of distributing P elements among N resonators, can be easily calculated using combinatory calculus techniques:

$$S = k\left\{\left(1 + \frac{U}{\varepsilon}\right)\ln\left(1 + \frac{U}{\varepsilon}\right) - \frac{U}{\varepsilon}\ln\frac{U}{\varepsilon}\right\} \tag{20}$$

This formula was the same as (18): now the definite quantity of energy β finds a physical meaning, becoming the "element of energy" ε, while α is replaced by the new universal constant k, hence assuring the presence of one of the two expected universal constants.

1	2	3	4	5	6	7	8	9	10
7	38	11	0	9	2	20	4	4	5

Fig. 4 Example of one the possible complexions when we assume $N = 10$ and $P = 100$ (Planck 1901, p. 557)

At this point, Planck demonstrates his formulation of Wien's displacement law in terms of entropy, i.e., $S = f(U/\nu)$. This represents the third cornerstone in Planck's final derivation of the Kirchhoff universal function, after (1) and (5). The consequence is that it must be $\varepsilon = h\nu$, that is a dependence of energy on frequency never seen before in classical physics, with h a universal constant, and Eq. 20 becomes

$$S = k\left\{\left(1 + \frac{U}{h\nu}\right)\ln\left(1 + \frac{U}{h\nu}\right) - \frac{U}{h\nu}\ln\frac{U}{h\nu}\right\} \tag{21}$$

hence giving the final functional dependence of $S(U, \nu)$, again depending on two *universal* constants.

The black-body spectral emissivity is then found from (1) and (5) following the same reasoning utilized in 1899:

$$u_\nu = \frac{8\pi h\nu^3}{c^3}\beta\frac{1}{e^{h\nu/kT} - 1} \tag{22}$$

Finally, in continuity with the papers of November 1899 and January 1900, Planck ends up calculating the values of the new absolute constants h and k: $h = 6.55 \times 10^{-27}$ erg s *and* $k = 1.346 \times 10^{-16}$ erg/K.

As for the "natural units of measurements", after their very first introduction in November 1899 and January 1900, Planck re-proposed them in his book in 1906 (Planck 1906), when his theory begun to be accepted, with h in place of b and h/k in place of a. In fact, it was shown elsewhere by one of us [NR] that Planck's interest in universal constants, that he repeatedly put forward also throughout all the following editions of his book, provides a unifying key for reconstructing and interpreting his work (Robotti 2000; Robotti and Badino 2000, 2001).

The "natural units" defined in 1906 had the following values:

$$\text{unit of length} = \left(\frac{Gh}{c^3}\right)^{1/2} = 4.03 \times 10^{-33}\ \text{cm}$$

$$\text{unit of mass} = \left(\frac{ch}{G}\right)^{1/2} = 5.42 \times 10^{-5}\ \text{g}$$

$$\text{unit of time} = \left(\frac{Gh}{c^5}\right)^{1/2} = 1.34 \times 10^{-43}\ \text{s}$$

$$\text{unit of temperature} = \frac{1}{k}\left(\frac{c^3h}{G}\right)^{1/2} = 3.63 \times 10^{32}\ {}^\circ\text{C}$$

Initially met with disinterest, Planck's idea was eventually revived and refined from the 1930s onward with the advancement of the theories of general relativity and quantum mechanics. Today, Planck's "units of measurement" for mass, time, and length (using $\hbar$ instead of h) have the following values (where m_P, t_P, and l_P are the *Planck mass, Planck time,* and *Planck length,* respectively):

$$m_P = \left(\frac{\hbar c}{G}\right)^{1/2} = 2.18 \times 10^{-5}\ \text{g}$$

$$t_P = \left(\frac{\hbar G}{c^5}\right)^{1/2} = 5.31 \times 10^{-44}\ \text{s}$$

$$l_P = \left(\frac{\hbar G}{c^3}\right)^{1/2} = 1.62 \times 10^{-35}\ \text{m}$$

and they are aptly named "Planck values", reflecting their historical significance.

5 Conclusions

We presented in a didactical perspective at the higher education level a reconstruction based on the existing historical literature that brings out the historical and conceptual mistakes present in most common standardized introductions to the black-body spectrum and to the birth of quantum physics, and the great didactical value of following actual Planck's reasoning.

Although, as pointed out elsewhere (Parisi 2001, 2008), it was later recognized that the properties of matter are irrelevant to explain the black-body radiation spectrum, Planck's approach, based on the radiation–matter interaction and on calculating the entropy of the material resonators, gave the correct black-body radiation formula and opened the way to the birth of Quantum Physics.

Our analysis illustrates the main aspects that should not be neglected and should rather be emphasized in a correct didactic path on this subject, namely, the central role of Kirchhoff's universal law that relates electromagnetism and thermodynamics stating that the ratio between the emissivity and the absorptivity of electromagnetic radiation (i.e., the emissivity of the black body) for all bodies at any frequency depends only on the temperature; the work done by Wien, who theoretically (*not* experimentally) deduced the displacement law, the dependence of the black-body spectrum from the (ν/T) ratio, and the first expression for the temperature and wavelength dependence of Kirchhoff's law; Planck's objective of deriving this law from first principles (*not* looking for a good fit of an experimental curve), i.e., from the fundamental laws of electromagnetism, hence the role of the Hertz resonator model (*not* of the stationary modes inside a cavity), and from the second law of thermodynamics with the classical definition of the absolute thermodynamic temperature of a physical system in terms of its entropy and internal energy; the need to make use of the probabilistic interpretation of entropy, proposed by Boltzmann, for which Planck gave the generalized universal formulation now called "the Boltzmann law", leading to the Boltzmann constant k; the theoretically based recognition of the need for a finite element of energy and of its proportionality to the frequency, leading to the Planck constant, *h*; the profound link between quantum physics and thermodynamics, especially with reference to the second law of thermodynamics and to irreversibility; the role of the absolute universal constants, representing a unifying key for reconstructing and interpreting Planck's work; and Planck's proposal for a system of "natural units of measurement" and its role in modern developments of elementary particles theories and cosmology.

Moreover, Planck's route highlights some didactically relevant aspects related to the *nature of science*: the interplay between theory and experimental measurements; the interweaving between mathematics and physics; and the role of philosophical motivations such as the search for the absolute in the development of science. For these reasons, we believe that the historical-didactic path proposed in the present work will be a useful reference for improving the learning and teaching of both the birth of quantum physics and thermodynamics especially for the higher education level.

References

Badino M (2015) The bumpy road: max planck from radiation theory to the quantum (1896–1906). Springer, Cham
Boltzmann L (1877) Sitzungsberichte D. Akad. D. Wiss. Wien 76:373
Boltzmann L (1884) Ann der Physik 22:291
Branchetti L, Cattabriga A, Levrini O (2019) Phys Rev Phys Ed Res 15:020130
Darrigol O (1992) From c-numbers to q-numbers: the classical analogy in the history of quantum theory. University of California Press, Berkeley
Dougal RC (1976) Phys Educ 11:438
Duncan A, Janssen M (2019) In: Constructing quantum mechanics: volume 1: the scaffold: 1900–1923. OUP, Oxford
Gearhart CA (2002) Phys Perspect 4:170
Halliday D, Resnick R, Walker J (2021) Fundamental of physics, 12th edn. Chichester
Jeans JH (1905) Proc Roy Soc London A 76:545
Kirchoff GR (1860) Ann der Physik 109:275
Klein MJ (1962) Arch Hist Exact Sci 1(5):459
Lummer O, Pringsheim E (1899a) Verh der d. Phys Ges 1:23
Lummer O, Pringsheim E (1899b) Verh der d. Phys Ges 1:215
Lummer O, Pringsheim E (1900) Verh der d. Phys Ges 2:163
Parisi G (2001) In: Max Planck. L'inizio della nuova fisica. Accademia dei Lincei, Rome
Parisi G (2008) Planck's legacy to statistical mechanics (2008). https://arxiv.org/pdf/cond-mat/0101293
Paschen F (1896) Ann der Physik 58:455
Paschen F (1899a) Sitzungsberichte d. Akad. d. Wiss. Berlin 1:405; Astrophys J 10:40
Paschen F (1899b) Sitzungsberichte d. Akad. d. Wiss. Berlin 2:959; Astrophys J 11:288 (1900)
Planck M (1895) Ann der Physik 291(5):220
Planck M (1896) Ann der Physik 293(1):1
Planck M (1897a) Sitzungsberichte D. Akad. Wiss. Berlin 6:57
Planck M (1897b) Sitzungsberichte D. Akad. Wiss. Berlin 34:715
Planck M (1897c) Sitzungsberichte D. Akad. Wiss. Berlin 52:1122
Planck M (1898) Sitzungsberichte D. Akad. Wiss. Berlin 34:449
Planck M (1899) Sitzungsberichte D. Akad. Wiss. Berlin 23:440

Planck M (1900a) Ann der Physik 306(1):69
Planck M (1900b) Ann der Physik 306(4):719
Planck M (1900c) Verh der d. Phys Ges 2(13):202
Planck M (1900d) Verh der d. Phys Ges 2(17):237
Planck M (1901) Ann. der Physik 309(3):553
Planck M (1906) Vorlesungen über die Theorie der Wärmestrahlung. Barth, Leipzig
Planck M (1920) The genesis and present state of development of the quantum theory. Nobel Lecture. http://www.nobelprize.org/nobel_prizes/physics/laureates/1918/planck-lecture.html
Planck M (1948) Wissenschaftliche Selbstbiographie. Barth, Leipzig
Planck M (1972) In: Kangro H (ed) Original papers in quantum physics. Taylor & Francis, London
Rayleigh JWS (1900) Phil Mag 49:539
Robotti N (2001) In: Beltrametti E, Giuliani G, Rimini A, Robotti N (eds) One Hundred Years of h, SIF, Bologna, p 1
Robotti N, Badino M (2000) In: The foundations of quantum mechanics. World Scientific, Singapore, p 343
Robotti N, Badino M (2001) Ann Sci 58:137
Rosenfeld L (1936) Osiris 2:149
Rubens H, Kurlbaum F (1900) Verh der D. Phys Ges 2:929; Astrophys J 14:335 (1901)
Serway RA, Jewett JW (2019) Physics, 10th edn. Cengage, Boston
Stefan J (1879) Sitzungsberichte D. Akad. D. Wiss. Wien 79:391
Walker J (2021) Physics, 5th edn. Pearson
Wien W (1893) Sitzungsberichte d. Akad. d. Wiss. Berlin, 55
Wien W (1896) Ann der Physik 58:662
Zettili N (2022) Quantum mechanics: concepts and applications, 3rd edn. UK, Wiley, Chichester

Engagement of Student in Learning Physics

The Role of Academic Engagement on Freshmen's Performance in an Introductory Physics Course

Agostino Cioffi, Silvia Galano, and Italo Testa

Abstract In this paper, we aim to understand the relationship between academic engagement and students' performance in the first university year, specifically in an introductory physics exam in the first semester of an engineering course. To this aim, we adopted the model of engagement developed by Freda et al. and explored if the engagement dimensions have a differential effect on students' performance controlling also for high school background. Overall, 134 first-year university students (female students = 32.8%) participated in the study. A binary logistic regression was used to analyze data. Results show that among the engagement dimensions, peer interaction and university and relational network significantly affected the passing of the physics exam. However, the two dimensions have contrasting effects on students' performance. Our results suggest that the university context in which students find themselves can be more relevant for their performance with respect to the perceived value of chosen degree course. Moreover, specific exams may present more difficulties to students who need external support (e.g., parents, friends). Implications for higher education policies are briefly discussed.

1 Introduction

Higher education plays a crucial role in developing social resources for individuals (Attanasio and Porcu 2021). The university system is responsible for implementing inclusive processes to ensure that more students have access to and benefit from this resource (Barone 2011). European Union has allocated significant funds to promote guidance in primary, secondary, and tertiary education, making student inclusion a central issue especially for higher education institutions (Macarie and Moldovan 2015). To ensure effective decision-making processes on behalf of the students, it is necessary to harmonize school, university, and the labor market, especially in the Science–Technology–Engineering–Mathematics area (STEM) (Miller and Wai 2015). The degree courses in this area are male dominated and have high

A. Cioffi · S. Galano · I. Testa (✉)
Department of Physics E. Pancini, University of Naples Federico II, Naples, Italy
e-mail: italo.testa@unina.it

I. Testa et al. (eds.), *Connecting Physics Education Research and Practice*,
Challenges in Physics Education, https://doi.org/10.1007/978-3-031-86609-8_12

dropout rates, particularly in the first year of enrollment (Ulriksen et al. 2015). Therefore, universities should aim to comprehend the strengths and limitations of various profiles of undergraduate students and guide them throughout their academic journey since the first year on enrollment (Salmieri and Giancola 2020). For instance, some students can complete their university studies and earn their degree without difficulty by relying on their own resources, whether individual or social. However, others may experience delays in their academic career or encounter various challenges, such as performance-related or relational issues (Modena et al. 2021). Finally, some students may choose to drop out of university when faced with the different challenges that university life presents (Belloc et al. 2010).

One of the most concerning data is the high rate of university dropouts. In 2016, approximately 3 million students dropped out of university in Europe alone (Aina and Pastore 2020; Eurostat 2016). The situation is particularly alarming in Italy, where the dropout rate was around 14.5% in 2018, with higher percentages in the South (18.8%), followed by the North (12.2%) and the Center (10.7%) (ISTAT 2018). Despite access to university study programs is granted for all students with an upper secondary school diploma and regulated by entrance exams only in fields like medicine, nursey, and architecture, recent national data shows that only 20.1% of Italy's population aged 25–64 holds a degree, compared to the EU's 32.8% (ISTAT 2021). Additionally, in 2020, at a national level only 24.9% of graduate students obtained a STEM degree (Attanasio et al. 2018). Therefore, addressing this issue should be a priority for the Italian university system at both national and local levels.

The issue of university dropout is also important since it affects individuals and society as a whole. For instance, individuals who do not complete college have poorer overall health in later life and are more likely to be unemployed (Rumberger 2011). They are also at risk of developing various psychopathologies, such as depression and anxiety disorders that could interfere with their educational and developmental tasks (Beiter et al. 2015; Deroma 2009; Adlaf et al. 2001; Bewick et al. 2010; Grøtan et al. 2019). Structurally, university dropouts are costly for the economy due to unemployment, public health expenditures, and declining tax contributions (Tyler and Lofstrom 2009; Freudenberg and Ruglis 2007).

Previous research has indicated that first-year students have high dropout rates (Contini et al. 2018). This is a critical period for many students as they transition from secondary school to university and may struggle to cope with the inherent difficulties (Wintre and Bowers 2007). To ensure effective monitoring of dropout intentions, it is necessary to begin from the first year of the academic career and analyze possible trajectories in order to identify strengths and limitations and enable the development of ad hoc prevention interventions (Brito et al. 2019). Dropout can be influenced by various socio-demographic and contextual variables (Aina et al. 2022). For instance, students who come from disadvantaged socio-economic backgrounds are female or are first-generation students (i.e., the first in their family to enroll in a university course) are more likely to experience poor academic performance and, ultimately, dropout of university (Caprile et al. 2015). While acknowledging the important role played by these variables in early dropout, however, this study aims to investigate the affective factors that influence the risk of early school leaving, specifically academic engagement.

Engagement in the higher education literature has typically been defined as a multidimensional construct that measures the intensity of a student's involvement in a given activity and that includes the following components: cognitive, emotional, and behavioral (Fredricks and McColskey 2012). Wang and colleagues developed a model with a further dimension to measure social engagement in science and mathematics (Fredricks et al. 2016), while Reeve and Tseng included a dimension to account for proactive engagement (Reeve and Tseng 2011). Given the domain-specific nature of engagement, in these previous conceptualizations, the cognitive and behavioral dimensions often overlap with self-regulation and metacognition (Cera et al. 2013; Manganelli et al. 2015), thus contributing to making the definition of the engagement construct problematic (Sinatra et al. 2015; Corno and Mandinach 1983).

In this work, we will refer to the model of engagement developed by Freda and colleagues at our university (Freda et al. 2023). The model includes six dimensions: (1) the value of the university and sense of belonging; (2) the value of the university course; (3) the perception of the ability to persist in the university choice; (4) the relationship with university professors; (5) the relationship with university peers; and (6) the relationship between the university and the relational network. The first two dimensions relate to the continuous development of skills and competences in the awareness and reflective use of the academic learning relationship, and to the ability to position oneself in the university context in a manner consistent with one's life professional goals. The third dimension relates to the intention to continue attending the university program despite the difficulties and obstacles that may be encountered throughout the academic trajectory. The fourth and fifth dimensions concern the social relations built during the study program with teachers and colleagues, which in turn affect the sense of belonging and connection to the academic institution. Finally, the sixth dimension refers to the integration of the university project and its value outside the university context and in one's broader life (e.g., in the family context) (Girelli et al. 2018; Kahu 2011).

2 Research Question

While the literature in science education has shown that engagement is an important construct in predicting student persistence in university choice, previous studies have focused on business and economics, with little focus on STEM courses. In particular, there is a lack of studies on first-year undergraduate STEM courses, where dropout rates are among the highest. In this paper, we focus on an engineering course with an introductory physics class taught in the first semester. The reason for this choice is that the introductory physics class is usually perceived to be difficult and failing it may lead to an increased likelihood of dropping out later in the academic year. Thus, the following research questions guided the study:

To what extent is academic engagement related to the performance in an introductory physics exam in the first semester of an engineering course, after controlling for high school background?

3 Methods

3.1 Sample and Instructional Context

This study involved $N = 134$ first-year university students (female students = 32.8%) enrolled in the following engineering courses: Automation (10.0%), Electronics (10.0%), Computer Science (46.0%), and Biomedical Engineering (34.0%). For all these courses, there is no mandatory entrance exam, but only a self-assessment test. The students attended the first semester for a total period of about 3 months, during which they took the following courses: Calculus I (72 h), Introductory Physics (48 h), and Programming (72 h). About half of the students (45.5%) had attended a mathematics- and science-oriented high school, about a fifth (20.9%) had attended a high school for applied sciences, a quarter (26.1%) had attended a technical high school, while the remaining students (7.5%) had attended other types of high schools. As for the final school grade, which in Italy ranges from 60 to 100, about one-fifth of the students (19.4%) received a grade between 60 and 89, 39.6% of the students received a grade between 90 and 99, while the remaining 41.0% of the sample received a grade of 100.

The introductory physics course follows the standard university syllabus. In particular, the student is expected to acquire the basic concepts of classical mechanics and the first concepts of thermodynamics, emphasizing the methodological and phenomenological aspects. In addition, students are expected to acquire skills in solving simple and complex physics problems. The course syllabus includes kinematics of the material point in one dimension, principles of dynamics, work and kinetic energy, conservation of mechanical energy and momentum, angular momentum and moment of force, dynamics of rigid bodies, rotation about a fixed axis, elements of statics and fluid dynamics, temperature and heat, first law of thermodynamics, ideal gases, and thermodynamic transformations.

3.2 Instruments and Measures

To measure student engagement, the Student Academic Engagement Scale (SAES) questionnaire validated in (Freda et al. 2023) for Italian first-year students of STEM degree courses was used. The instrument consists of 29 items aimed at measuring the six described dimensions of academic engagement on a five-level scale of agreement (1 = not at all agree; 5 = strongly agree). We first verified that the sample was large enough to carry out an exploratory factor analysis (KMO sample adequacy test =

Table 1 Factors of SAES questionnaire and retained items after exploratory factor analysis

Factor	Items
1. Peer interaction	I have good relationships with my university colleagues I've made meaningful friends with some college colleagues I like to meet friends at university I feel like I'm part of a group of friends at university
2. Intention to drop-out	I'd better do other things than go to university I'd leave university right away if I had an alternative Sometimes I think about leaving university I am convinced that my decision to enroll in this university was the right one for me (reversed) I'm sure the degree course I'm attending is the right place for me (reversed)
3. Interaction with instructors	Teachers clarify what they expect of us students Teachers are usually available to discuss my work My teachers respect me as a person My teachers are interested in my opinions and what I say
4. Usefulness of the course of study	The course of study I'm attending is functional to the achievement of my professional goals The course of study I'm attending is interesting I like the course of study I'm attending I think university is good for me
5. University and relational network	I talk about my professional plans with my friends I talk about my professional plans with my family I discuss with my family about my university path

0.689, Bartlett sphericity test = $p << 0.01$). Then, we ran the exploratory factor analysis, which showed the presence of five factors that explained a cumulative total variance of 57.0%. Overall, 20 items were retained. Note also that the original six-factor structure was not confirmed in the present study. In particular, the new Factor 4 corresponds to the two original factors: value of the university and sense of belonging; and value of the university course. Table 1 reports the extracted factors and the corresponding items.

Table 2 summarizes for each factor of the descriptive statistics, the reliability coefficient (Cronbach's alpha), and the correlations with the other factors.

3.3 Data Analysis

We used the logistic regression to answer our research question. In its simplest form, when there is only one predictor variable X_1, the logistic regression equation is simply given by

Table 2 Descriptive statistics, reliability coefficient, and correlations between engagement factors

	1	2	3	4	5
1. Peer interaction	0.83[a]				
2. Intention to drop-out (reversed)	−0.105	0.75[a]			
3. Interaction with instructors	−0.017	−0.372**	0.77[a]		
4. Usefulness of the course of study	0.144	−0.164*	0.050	0.69[a]	
5. University and relational network	0.256**	−0.167*	0.073	0.122	0.76[a]
Mean	3.6	2.0	3.3	4.2	3.4
Standard deviation	0.9	0.7	0.9	0.6	1.0
Asymmetry	−0.635	0.737	0.039	0.841	−0.286
Kurtosis	−0.298	0.168	−0.519	−0.863	0.632

[a] Cronbach's Alpha; * The correlation is significant at the 0.05 level (two sided); ** The correlation is significant at the 0.01 level (two sided)

$$P(Y) = \frac{1}{1 + \exp[-(b_0 + b_1 X_1)]} \tag{1}$$

where $P(Y)$ is the probability that Y occurs, i.e., the probability that a case belongs to a certain category, and b_0 and b_1 are two constants to be determined. The resulting value of the equation varies between 0 and 1. A value close to 0 means that Y is very unlikely to have occurred, while a value close to 1 means that Y is very likely to have occurred. Just like linear regression, one can extend this equation to include multiple predictors. When there are multiple predictors, the equation becomes

$$P(Y) = \frac{1}{1 + \exp[-(b_0 + b_1 X_1 + b_2 X_2 + \ldots + b_n X_n)]} \tag{2}$$

Similarly, as in linear regression, each predictor variable in the logistic regression equation has its own associated parameter (b), which is estimated from the sample data. While for linear regression these parameters are estimated using the least squares method, in logistic regression the maximum likelihood estimate is used (Berry 1993; Tabachnick and Fidell 2012). The maximum likelihood estimate statistic is analogous to the residual sum of squares of multiple regression in the sense that it is an indicator of how much "unexplainable" information there is in the model. A crucial parameter in the interpretation of logistic regression is the odds ratio, which is represented by $\exp(B)$ where B is the standardized regression coefficient corresponding to a given predictor. When the predictor variable is categorical, such as a binary variable that assume the values of 0–1, this parameter represents the ratio of the probability of an event occurring when the predictor variable is 1 to the probability of an event occurring when the variable is 0.

In this study, the passing of the physics exam was used as a binary dependent variable (exam failed = 0; exam passed = 1), the five engagement dimensions and school background (type of school attended and performance at the final exam) were used as independent variables. We acknowledge that using a dichotomous outcome variable as a proxy for the performance hampers the depth of the analysis since the performance construct is more fine-grained than passing the exam or not.

4 Results

Table 3 shows the results of the logistic regression. The results show a significant effect of the high school final examination grade, the type of school attended, the peer interaction and the university and relational network factor of engagement. The other dimensions of engagement are not significantly associated with success in the general physics exam.

The relationships between independent variables were also investigated. We first explored whether the high school final examination grade was associated with student engagement in the physics course. To this aim, we first categorized the students according to their high school final examination grade and then studied the differences of the average scores in the engagement dimension through a one-way analysis of variance (ANOVA). Based on the distribution of high school final examination grade, students were divided in three groups: 60 < grade < 89, 90 < grade < 99, and grade = 100. The reason for the choice of these intervals was to obtain groups of students consisting of approximately the same number.

The distribution of the scores in the five dimensions of engagement for the three groups is shown in Table 4. The ANOVA shows a slight significant difference only for the peer interaction dimension of engagement, $F(2; 131) = 3.061; p = 0.050$. Second, we investigated whether the type of high school attended has influenced engagement in the physics course. Given the distribution across the different types of high school,

Table 3 Results of the logistic regression

Independent variable	B	S.E	Gl	Odds ratio	95% C.I. per odds ratio	
					Inferior	Superior
Peer interaction	0.545*	0.228	1	1.699	1.086	2.658
University and relational network	–0.529*	0.208	1	0.667	0.444	1.003
State exam grade = 100	1.401***	0.399	1	4.534	2.073	9.918
High school = Scientific	0.832*	0.421	1	2.298	1.074	6.937

$R^2 = 0.195$ (Cox & Snell), 0.263 (Nagelkerke). Model $\chi^2(4) = 29.091, p < 0.001$; * $p < 0.05$, ** $p < 0.01$, *** $p < 0.001$

Table 4 Average scores in the engagement dimensions for the three groups of students according to high school final examination grade

Factor	Group	N	Average	S.D
1. Peer interaction	Grade between 60 and 89	26	3.4	0.9
	Gade between 90 and 99	53	3.6	0.9
	Grade = 100	55	3.9	0.9
	Total	134	3.7	0.9
2. Intention to drop-out (reversed)	Grade between 60 and 89	26	2.1	0.7
	Rating between 90 and 99	53	2.0	0.7
	Rating = 100	55	1.9	0.6
	Total	134	2.0	0.7
3. Interaction with instructors	Grade between 60 and 89	26	3.2	0.7
	Rating between 90 and 99	53	3.4	0.9
	Rating = 100	55	3.3	0.9
	Total	134	3.3	0.9
4. Usefulness of the course of study	Grade between 60 and 89	26	4.3	0.6
	Rating between 90 and 99	53	4.2	0.6
	Rating = 100	55	4.2	0.6
	Total	134	4.2	0.6
5. University and relational network	Grade between 60 and 89	26	3.5	1.2
	Rating between 90 and 99	53	3.3	1.1
	Rating = 100	55	3.6	0.8
	Total	134	3.0	1.0

the students who had followed the math- and science-oriented high school were coded with 1 while the students who had attended other types of high schools with 0. Then, the average scores in each of the five dimensions of engagement of the two groups were compared by means of a t-test. The results are shown in Table 5. We observe that the students who have followed a math- and science-oriented high school have significantly lower scores than students who have followed other types of high school in the intention to drop-out (reversed) and the university and relational network dimensions.

5 Discussion and Conclusions

First of all, our study confirms that academic engagement has a significant relationship with students' performance in the first semester, in particular, in the physics course (Passeggia et al. 2023). Among the five dimensions of engagement, peer interaction and university and relational network were the dimensions mostly related with

Table 5 Average scores in the engagement dimensions for the two groups of students according to the type of high school attended

Engagement dimension	Type of high school	N	Average	S.D	*t*	Cohen's d
1. Peer interaction	Math and science oriented	61	3.8	0.8	1.321	0.224
	Others	73	3.6	1.0		
2. Intention to drop-out (reversed)	Math and science oriented	61	1.9	0.6	–2.132*	–0.361
	Others	73	2.1	0.8		
3. Interaction with instructors	Math and science oriented	61	3.4	0.8	0.962	0.167
	Others	73	3.3	0.9		
4. Usefulness of the course of study	Math and science oriented	61	4.3	0.5	0.974	0.169
	Others	73	4.2	0.7		
5. University and relational network	Math and science oriented	61	3.8	0.9	3.525***	0.611
	Others	73	3.2	1.0		

* $p < 0.05$, ** $p < 0.01$, *** $p < 0.001$

the passing of the physics exam. Interestingly, the two dimensions have contrasting effects. In particular, students who perceived they have more interaction with their peers are more likely to pass the physics exam. On the other hand, students who feel the need of support in their university studies from familiar and personal relationships are more likely to fail the General Physics I exam in the first semester. This result suggests that it is necessary to improve the collaboration with peers so that they feel included in a network of relationships that helps them overcome the difficulties of the school–university transition (Priulla and Attanasio 2023). The significant role of engagement with peers with respect to, for example, the value given to the university suggests that the university context in which students find themselves can be more relevant for their performance with respect to the perceived value of chosen degree course. Conversely, the negative effect of the university and relational network on passing the physics exam seems to suggest that this specific exam may present more difficulties to students who need external support (e.g., parents, friends) when acting a certain behavior. Finally, our study shows that the grade received at the high school, in particular, the highest possible grade and having attended the math- and science-oriented high school are two of the factors that favor passing the Introductory Physics

I exam. This result implies that the first semester of university is not sufficient to smooth out the differences in students' previous preparations.

Overall, these results are limited by the small sample involved in the study. For this reason, the relationships explored in this paper will be further studied in an upcoming university project that aims to understand the role of academic engagement on student learning and dropout intention and to study through path analysis how secondary school data relates to engagement, and how engagement in turn affects academic performance.

References

Adlaf EM, Gliksman L, Demers A, Newton-Taylor B (2001) The prevalence of elevated psychological distress among Canadian undergraduates: findings from the 1998 Canadian campus survey. J Am Coll Heal 50(2):67–72. https://doi.org/10.1080/07448480109596009

Aina C, Baici E, Casalone G, Pastore F (2022) The determinants of university dropout: a review of the socio-economic literature. Soc-Econ Plann Sci 79. https://doi.org/10.1016/j.seps.2021.101102

Aina C, Pastore F (2020) Delayed graduation and overeducation in Italy: a test of the human capital model versus the screening hypothesis. Soc Indic Res 152:533–553. https://doi.org/10.1007/s11205-020-02446-0

Attanasio M, Priulla A, Albano A, Enea M (2018) Analisi delle carriere universitarie nelle lauree scientifiche di base in Italia nell'ultimo decennio. Induzioni: demografia, probabilità, statistica a scuola 57(2):37. https://doi.org/10.19272/201800902002

Attanasio M, Porcu M (2021) Rapporto sulla popolazione. L'Italia e le sfide della demografia, vol 793. Universale Paperbacks Il Mulino, p 163

Barone C (2011) Some things never change: gender segregation in higher education across eight nations and three decades. Soc Educ 84:157–176. https://doi.org/10.1177/0038040711402099

Beiter R, Nash R, McCrady M, Rhoades D, Linscomb M, Clarahan M et al (2015) The prevalence and correlates of depression, anxiety, and stress in a sample of college students. J Affect Disord 173:90

Belloc F, Maruotti A, Petrella L (2010) University drop-out: an Italian experience. High Educ 60:127–138. https://doi.org/10.1007/s10734-009-9290-1

Berry WD (1993) Understanding regression assumptions. Sage Publications Inc.

Bewick B, Koutsopoulou G, Miles J, Slaa E, Barkham M (2010) Changes in undergraduate students' psychological well-being as they progress through university. Std High Educ 35:633 (2010). https://doi.org/10.1080/03075070903216643

Brito M, Medeiros F, Bezerra EP (2019) An infographics-based tool for monitoring dropout risk on distance learning in higher education. In: 18th international conference on information technology based higher education and training (ITHET), Magdeburg, Germany, pp 1–7. https://doi.org/10.1109/ITHET46829.2019.8937361

Caprile M, Palmen R, Sanz P, Dente G (2015) Encouraging STEM studies labour market situation and comparison of practices targeted at young people in different member states. Pol Dep A 12. http://hdl.voced.edu.au/10707/371460

Cera R, Mancini M, Antonietti A (2013) Relationships between metacognition, self-efficacy and self-regulation in learning. J Educ Cult Psychol Std 7(7):115

Contini D, Cugnata F, Scagni A (2018) Social selection in higher education. Enrolment, dropout and timely degree attainment in Italy. High Educ 75:785–808. https://doi.org/10.1007/s10734-017-0170-9

Corno L, Mandinach EB (1983) The role of cognitive engagement in classroom learning and motivation. Educ Psychol 18(2):88. https://doi.org/10.1080/00461528309529266
Deroma L (2009) The relationship between depression and college academic performance. Coll Stud J 43(2):325–334
Eurostat (2016)
Freda MF, Raffaele DLP, Esposito G et al (2023) A new measure for the assessment of the university engagement: the SInAPSi academic engagement scale (SAES). Curr Psychol 42:9674. https://doi.org/10.1007/s12144-021-02189-2
Fredricks JA et al (2016) Using qualitative methods to develop a survey measure of math and science engagement. Learn Instr 43:5. https://doi.org/10.1016/j.learninstruc.2016.01.009
Fredricks JA, McColskey W (2012) The measurement of student engagement: a comparative analysis of various methods and student self-report instruments. In: Christenson SL, Reschly AL, Wylie C (eds) Handbook of research on student engagement. Springer Science+Business Media, pp 763–782. https://doi.org/10.1007/978-1-4614-2018-7_37
Freudenberg N, Ruglis J (2007) Reframing school dropout as a public health issue. Prev Chronic Dis 4(4)
Girelli et al (2018) Autonomy supportive contexts, autonomous motivation, and self-efficacy predict academic adjustment of first-year university students. Front Educ 3. https://doi.org/10.3389/feduc.2018.00095
Grøtan K, Sund ER, Bjerkeset O (2019) Mental health, academic self-efficacy and study progress among college students—The SHoT study Norway. Front Psychol 10:45. https://doi.org/10.3389/fpsyg.2019.00045
ISTAT (2018)
ISTAT (2021)
Kahu ER (2011) Framing student engagement in higher education. Std High Educ 38(5):758–773. https://doi.org/10.1080/03075079.2011.598505
Macarie FC, Moldovan O (2015) Horizontal and vertical gender segregation in higher education: EU 28 under scrutiny. Manag Challeng Contemp Soc 8:162–169
Manganelli et al (2015) Italian adolescents' civic engagement and open classroom climate: the mediating role of self-efficacy. J Appl Dev Psychol 41:8
Miller DI, Wai J (2015) The bachelor's to Ph.D. STEM pipeline no longer leaks more women than men: a 30-year analysis. Front Psych 6:37. https://doi.org/10.3389/fpsyg.2015.00037
Modena F, Rettore E, Tanzi GM (2021) Does gender matter? The effect of high performing peers on academic performances. The effect of high performing peers on academic performances. Bank of Italy Temi di Discussione (Working Paper) 1356. https://doi.org/10.2139/ssrn.4016597
Passeggia R et al (2023) Examining the relation between first-year university students' intention to drop-out and academic engagement: the role of motivation, subjective well-being and retrospective judgements of school experience. Innov High Educ 48:837–859. https://doi.org/10.1007/s10755-023-09674-5
Priulla A, Attanasio M (2023) Unveiling gender disparities in university pathways: insights from Italy's master's level. Euro J High Educ 1. https://doi.org/10.1080/21568235.2023.2247189
Reeve J, Tseng C-M (2011) Agency as a fourth aspect of students' engagement during learning activities. Cont Educ Psychol 36(4):257. https://doi.org/10.1016/j.cedpsych.2011.05.002
Rumberger R (2011) Dropping out: why students drop out of high school and what can be done about it. https://doi.org/10.4159/harvard.9780674063167
Salmieri L, Giancola O (2020) Gender differences in tertiary educational attainment and the intergenerational transmission of cultural capital in Italy. Educ Gend Gend Educ 77
Sinatra GM, Heddy BC, Lombardi D (2015) The challenges of defining and measuring student engagement in science. Educ Psychol 50(1):1. https://doi.org/10.1080/00461520.2014.1002924
Tabachnick BG, Fidell LS (2012) Using multivariate statistics, 6th edn. Pearson
Tyler JH, Lofstrom M (2009) Finishing high school: alternative pathways and dropout recovery. Fut Child 19(1):77. https://doi.org/10.1353/foc.0.0019

Ulriksen L, Madsen LM, Holmegaard HT (2015) Why do students in stem higher education programs drop/opt out? Explanations offered from research. In: Understanding student participation and choice in science and technology education. Springer, pp 203–217. https://doi.org/10.1007/978-94-007-7793-4_13

Wintre MG, Bowers CD (2007) Predictors of persistence to graduation: extending a model and data on the transition to university model. Can J Behav Sci 39:220. https://doi.org/10.1037/cjbs2007017

Orientation Activities for High School Students: Propaedeutic Test and Open-Ended Physics Lab

Roberto Virzi, Matteo Bozzi, Roberto Mazzola, and Maurizio Zani

Abstract As part of the project of Italian Ministry of University and Research related to active orientation from school to university, among the courses proposed by the Politecnico di Milano was the one titled "Methodological Introduction to Engineering Test Preparation". The aim of this course was to provide methodological tools for studying Physics and Mathematics topics and developing logical-mathematical skills, scientific methods, modelling, and problem-solving. The course was structured into modules covering motivational aspects, Mathematics and Physics. The latter consisted of an introductory online lesson on learning aspects, followed by an in-person laboratory activity. During the introductory session, students completed a multiple-choice questionnaire, aiming to highlight gaps or misconceptions that typically make the Engineering Test challenging. The laboratory activity was designed following an open-ended lab model. Students were encouraged to explore, with maximum autonomy, the physical quantities related to a certain phenomenon and discover the relationships between them. The research activity reported here focuses on validating the preparatory questionnaire.

1 Introduction

During the academic year 2022/2023, the Politecnico di Milano offered a wide range of orientation courses for high school students. These courses were designed in accordance with the guidelines outlined in the National Recovery and Resilience Plan (PNRR) (Decreto ministeriale 2022), ensuring specific criteria in terms of duration and content (Corsi di orientamento attivo nella transizione scuola università 2023). To be more specific, these courses were structured to span 15 h, aiming primarily to assist students in exploring and comprehending the university environment while fostering awareness of their expectations and aspirations for future studies. Among these offerings was our orientation course titled *Introduzione metodologica*

R. Virzi (✉) · M. Bozzi · R. Mazzola · M. Zani
Department of Physics, Politecnico di Milano, Milan, Italy
e-mail: roberto.virzi@polimi.it

I. Testa et al. (eds.), *Connecting Physics Education Research and Practice*, Challenges in Physics Education, https://doi.org/10.1007/978-3-031-86609-8_13

alla preparazione al test di Ingegneria (Methodological Introduction to Engineering Test Preparation). As implied by its name, the course targeted high school students seeking to equip themselves for the Engineering Test named *Test On Line* (TOL) (What the test consists of and how to prepare 2024) required for admission to the Engineering faculties at Politecnico di Milano University.

The course was structured in different moments. 3 h were dedicated to an in-presence lesson focused on motivation. The remaining 12 h were equally divided into more specific activities of Physics and Mathematics. Our group provided the Physics part organizing 2 h of an online introductory lesson during which students were asked to answer to an 8-item propaedeutical physics test. After that, students were asked to participate in a 4 h laboratory session. In total, 113 students have joined the proposal. The whole course was conducted in Italian.

At the end of this experience, we had the possibility to analyze all the data of participants applying the Classical Test Theory (Ding and Beichner 2009) in order to validate our questionnaire and the effectiveness of distractors. The following sections presented all the results of these analyses and also some consideration on how we could improve the test for future editions according to these results.

All these results are presented in the following sections to give an answer to the research question: is our test reliable and useful for the purposes of our orientation course?

2 The Open-Ended Style Laboratory

This work is particularly focused on the analysis of the propaedeutical test but before to go deeper in this topic it's useful to know something more about the structure of the laboratory. The main aim of our laboratory activity was to help students in developing skills in order to be able to discover and study physical contents in an autonomous way. These aims could be achieved, for instance, by asking students to discuss in their group how to solve an issue they found during their work or how to choose which quantities are useful and what should be a proper method to obtain a reliable measure. Considering this we preferred to avoid a classical approach and choose an open-ended style (Trumper 2003). This means that students did not receive detailed instructions about how to perform an experiment but only a very general description of the phenomenon they were supposed to investigate and some suggestions concerning which physical quantities they should consider. Each group (normally no more than 4 people per group) had to work on a different topic and had access to very simple material to set up the experiment. During the laboratory session, students could rely on the support of our research group which was composed of an associated professor, a research fellow, a Ph.D. candidate, and a graduating student. At the end of these 4 h, each group was asked to present their work to the rest of the class. This different approach has a positive influence because it allows students to feel more like protagonists of their work and makes them aware of the experimental nature of Physics (Wilcox and Lewandowski 2016). Moreover, it enhances the development of critical thinking and fosters the science self-efficacy. Self-efficacy is related to how a person believes in his capability to perform a task (Hu et al. 2022). In particular, science self-efficacy is very important for the learning process because it is strictly related to science interest and science achievement (Alhadabi 2021) that are fundamental for an orientation activity. It has also been observed that self-efficacy can help students to cope with more challenging tasks with less stress (Bandura 1997). Recent studies also highlighted that self-efficacy can affect the results of students in Physics exams (Malespina and Singh 2024). In addition, we think that this methodology can stimulate student's curiosity and help them to develop an interest for science knowledge and science learning that could be important in their future path (Hazari et al. 2010).

To give a proper idea of how our activities were organized we provide here a brief explanation of a laboratory session. First, students were divided into small groups composed by 3, 4, or a maximum of 5 persons per group. Then, each group received a box with different materials and a card containing very general information about the phenomena they were asked to study. An example of a card is shown in Fig. 1.

We prepared 7 cards covering the following topics: motion and friction, oscillations, pendulum, motion in a viscous medium, density and volume, optics, and mechanical energy conservation. Then the teacher gave a brief introduction and after that students could start their work. During the lab, teachers continuously stimulated students to find solutions and invent strategies: these dialogs were useful for giving and receiving feedback. In the end, each group of students presented the results of their work, giving reasons for their choices and answering possible questions from teachers or coursemates.

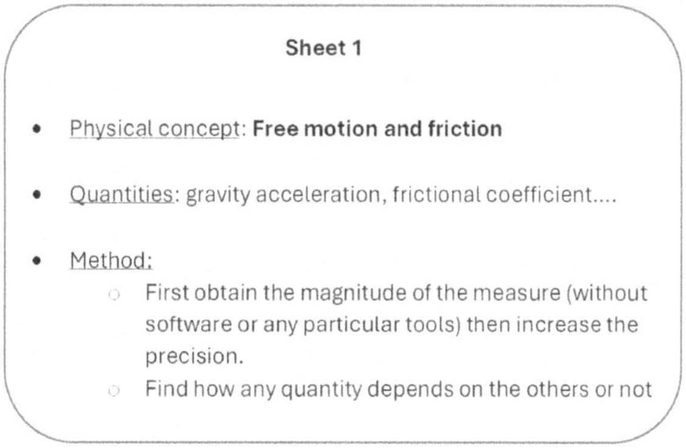

Fig. 1 Example of a card containing initial information to perform the laboratory experience

3 The Propaedeutical Questionnaire

The propaedeutical questionnaire that students faced in the introductory online lesson was useful to help them become aware of their gaps in the knowledge of Physics contents. The 8 questions forming our test were designed carefully (Ebel and Frisbie 1972) and share conceptual similarities, for example, with those found in the CSEM test (Maloney et al. 2001) or in the FCI (Hestenes et al. 1992). Each item follows a structure with 4 possible answers, wherein only one is correct, while the other 3 are intentionally designed to identify potential gaps or misconceptions. The topics covered by the 8 items are summarized in Table 1 .

In this questionnaire, each participant received 1 point for each correct answer and 0 point for each wrong or missing answer.

Furthermore, after completing the test, the instructor who conducted the introductory lesson provided the correct answers, accompanied by comments on incorrect choices, and sought feedback from participants. This aspect was important in prompting students to reflect on their knowledge and learning methods, fostering a meta-cognitive approach (Romainville 1994).

According to Hattie and Clarke (2019) feedback from teachers is fundamental in learning process especially when help students providing a direction to follow for their improvement. In the opposite direction, feedback from students can make the learning visible by teachers. In our situation, the feedback given by the instructor was focused on the misconceptions highlighted by the test in order to help students to develop skills useful for recognizing and avoiding them in the future. As for the feedback from students, given that the course was at the beginning, it was not related to what the students learned, but it provided useful information about the reasons for choosing a particular distractor. In this case, feedback could help students become aware of their mistakes and conscious of the way they learn and memorize.

Considering that the introductory meeting was online, this questionnaire was submitted using the Socrative platform (https://www.socrative.com/).

Table 1 List of topics covered by each item

Item	Topic
Q1	Force and movement
Q2	Friction
Q3	Circular motion
Q4	Kinetic and potential energy
Q5	Electric field and electric potential
Q6	Period and angular velocity
Q7	Free fall
Q8	Electrostatic force and Newton's third law

4 Test Validation

The validation is a fundamental step to be sure that our test is actually useful for our purposes (Bandalos 2018) or if it needs some changes. In our evaluation, we referred to several different coefficients that can provide useful information about single questions and the whole test as summarized by Ding and Beichner (2009). We could rely on data related to 113 students coming from 11 different schools and overall from 13 different school types.

Above all, we can observe in Fig. 2 that only very few students reached a score greater than the 50% of the total while the most part of them collected between 1 and 4 points up to 8. The average score of the sample is 2.67. It is particularly low but we want to underline that the aim of this test was not to evaluate the knowledge of a single student but to obtain an overview of gaps and misconceptions of the whole group. We can also notice that no one answered correctly to all questions. In this section we'll show the values of each coefficient we considered in our analysis.

In Table 2 are reported the Difficulty Index (P), the Discrimination Index (D), and the Point Biserial Coefficient (r_{pbi}) that provide an indicator for each item.

The difficulty index (P) is defined as the ratio between the number of students that answered correctly to a question and the total amount of students:

$$P_n = \frac{N_1}{N}$$

this means that a high difficulty level states that a high number of students answered correctly. In other words, the higher the difficulty index, the easier the question.

The value of P is useful even to calculate the point biserial coefficient defined as:

$$r_{pbi} = \frac{\overline{X_1} - \overline{X_0}}{\sigma_x}\sqrt{P_n(1 - P_n)}$$

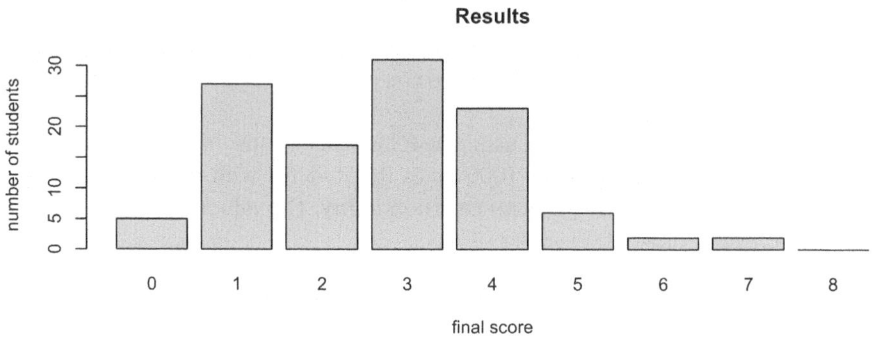

Fig. 2 Total score of the whole group of students

Table 2 Coefficient related to each item: difficulty index (*P*), discrimination index (*D*) and point biserial coefficient (r_{pbi})

Item	P	P-level (Crocker and Algina 1986)	D	D-level (Ebel and Frisbie 1972)	r_{pbi}
Q1	0.32	Medium	0.48	Good	0.45
Q2	0.38	Medium	0.56	Good	0.45
Q3	0.19	Medium-high	0.47	Good	0.51
Q4	0.28	Medium	0.58	Good	0.52
Q5	0.25	Medium-high	0.33	Reasonably good	0.30
Q6	0.50	Medium-low	0.50	Good	0.37
Q7	0.58	medium-low	0.55	Good	0.44
Q8	0.20	Medium-high	0.27	Marginal	0.27

where $\overline{X_1}$ is the average test score of students that answered correctly to a particular question, $\overline{X_0}$ is the average test score of students that answered wrong to the same question, σ_x is the standard deviation of scores and P is the difficulty level of the question. The point biserial coefficient expresses the reliability of a single question, in other words, it indicates the correlation between the item score and the total score. All our results can be labeled as *acceptable* because they are all above 0.2 (Kline 1986).

The Discrimination Index provides information about how much an item is able to recognize students with an overall high performance. To calculate it we used the following formula:

$$D_n = \frac{N_H - N_L}{N/4}$$

in which N_H is the number of students on the upper quartile that answered correctly to the question Q_n, N_L is the number of students in the lower quartile that answered correctly to the same question and N is the total number of students. Considering that our sample was composed of 113 persons and that the possible score ranged from 0 to 8 we found many situations of different students with the same score. This made hard to choose which student were to consider in the upper quartile and which in the lower quartile.

To fix this problem we decided, using the *R* language (https://www.r-project.org/) to write a script to sort randomly 1000 times the students with the same score and calculate the Discrimination Index for each possibility. The values reported in Table 2 are the average of these 1000 values of *D*.

Concerning *P* and *D* is reported in Table 2 also the level related to each value (Crocker and Algina 1986; Ebel and Frisbie 1972). As far as r_{pbi} concerns, all the values can be labeled as *acceptable* because they are all above 0.2 (Kline 1986). Observing the table we can notice that 3 questions had a medium-high difficulty ($P \leq 0.25$) while 2 had a medium-low difficulty ($0.75 \leq P < 0.50$). Concerning *D*

Table 3 Coefficient related to the whole test: Kuder-Richardson reliability (r_{test}) and Ferguson's delta (δ)

Coefficient	Value
r_{test}	0.32
δ	0.90

it's clear that all the items are good or reasonably good ($D \geq 0.30$) while only the Q8 was a marginal item ($0.20 \leq D < 0.30$). It's interesting to observe that Q3 has been a difficult question nevertheless it had a good discrimination index.

The last two coefficients we took in consideration were the Kuder-Richardson reliability (r_{test}) and Ferguson's delta (δ) as shown in Table 3 .

The Kuder-Richardson reliability is calculated with

$$r_{\text{test}} = \frac{K}{K-1}(1 - \frac{\Sigma P_i(1-P_i)}{{\sigma_x}^2})$$

where K is the number of item in the test, P_i is the difficulty level of question i and σ_x is the standard deviation of total scores. r_{test} expresses the internal consistency of the whole test.

Eventually, Ferguson's delta can be obtained with

$$\delta = \frac{N^2 - \Sigma {f_i}^2}{N^2 - N^2/(K+1)}$$

where N is the number of students, K is the number of items and f_i is the frequency of students with a total score i. δ indicates an overall discrimination level.

The value of r_{test} is 0.32 and it indicates that the test has low reliability and could be used for survey and group averages but not to assess a single student (Doran 1980). Nevertheless, as we wrote before, the aim of our test was exactly to underline gaps and misconceptions of the whole group of students so that a low Kuder-Richardson reliability was expected. In addition, r_{test} measures the internal coherence of a test, and remembering Table 1, it's clear that in a test in which each question covers a different topic it's hard to have a high internal coherence. On the contrary, the value of δ is high so our test provides a good level of discrimination among students (Kline 1986).

Eventually, we focused on the effectiveness of the distractors in each question. Table 4 shows all the percentages of choice for the four possible answers for each item. The percentage corresponding to the right answer is written in bold.

Observing the table it is possible to notice that only 3 distractors were chosen by <5% of the students so the most part of the distractors can be considered effective. More specifically, in 5 items there is a distractor that was chosen more often than the actual right answer, highlighting a very good effectiveness and suggesting the possible presence of a misconception.

According to these results, we made a general review of our test to be more effective for future editions of the course. In the next section, we show our changes.

Table 4 Percentages related to the choice of each answer. The percentage corresponding to the right answer is written in bold

Answer/item	Q1	Q2	Q3	Q4	Q5	Q6	Q7	Q8
A	**31.5**	42.0	62.8	19.6	33.6	21.2	5.4	66.1
B	9.9	2.7	**18.6**	12.5	20.9	**50.4**	33.9	8.0
C	49.5	**38.4**	2.7	**27.7**	**25.5**	8.8	**58.0**	6.3
D	9.0	17.0	15.9	40.2	20.0	19.5	2.7	**19.6**

5 Questionnaire Review

Taking into account the results obtained in Sect. 4 and the feedback received by students during the online lesson, we decided to make some modifications to Items 1, 2, 3, and 7.

Figure 3 shows the text of the first question. Remembering Tables 2 and 4 we can state that Q1 is generally a good item. Nevertheless, we observed that answers C and D are equivalent because the text does not specify if the stone is thrown from left to right or vice-versa. The most chosen answer was C and we think that the reason is that students probably interpreted the missing text as *from left to right* and applied the common misconception *motion implies active force* (Hestenes et al. 1992) or *impetus dissipation* (Hestenes et al. 1992). To avoid this possible misunderstanding we specified the thrown direction (left to right) and we changed the answers as shown in Fig. 4.

We expect that students who choose the answers B or D think that at the maximum point, to have zero vertical velocity, a zero net force is required.

Figure 5 shows the second item. In that case, we wanted to improve the answer B because it was a poor distractor chosen by less than 5% of students. The new answer B is "it does not depend on the applied force" so we also had to adjust the last

1. A boy throws a stone from the bank of a river to see if he can reach the other side. Which of the following diagrams best represents the system of forces acting on the stone (neglecting air resistance) at the moment it reaches the highest point in its trajectory?

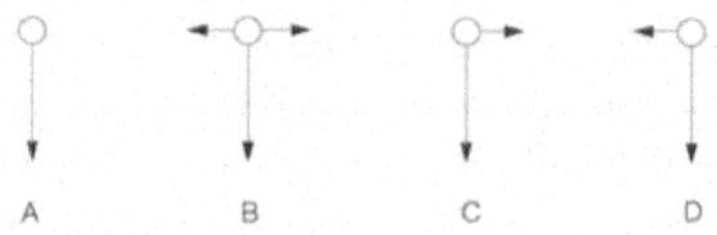

a) A
b) B
c) C
d) D

Fig. 3 Question 1 of our questionnaire. In bold the right answer

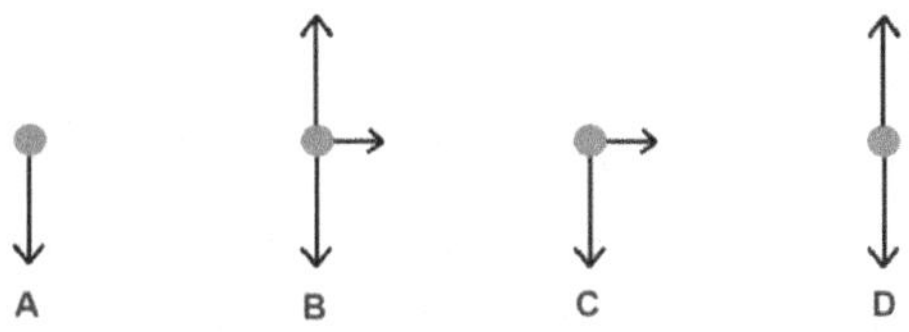

Fig. 4 New set of possible answers to item 1

Fig. 5 Question 2 of our questionnaire. In bold the right answer

2. Applying a horizontal force of F_1=5 N to an object placed on a horizontal surface, it is observed that the object does not move due to friction. Even when a horizontal force of F_2=10 N is applied, the object still does not move because of friction. What is the magnitude of the frictional force acting on the object?

a) It is greater than 10 N
b) It is equal to 10 N
c) **It is 5 N in the case of F_1 and 10 N in the case of F_2**
d) It cannot be determined with the given data

sentence of the item changing it to "which statement related to friction force acting on the object is correct?". We think that this new distractor could detect a common misconception about friction: *Friction is constant and equals to* $f = \mu N$ *(normal force)* (Liu and Fang 2016).

Even in the third item, shown in Fig. 6, we had a poor distractor because answer C was chosen only by 2.7% of students. We decided then to change it to "The centripetal acceleration is approximately 4m/s^2". This new formulation aims to detect students who are aware that there is an acceleration greater than zero but who calculate it wrongly by dividing speed by the period.

Eventually, we focused on item 7 to improve the effectiveness of distractors A and D . Starting from the item shown in Fig. 7, we changed answer A to "It varies during the fall" and answer D to "It depends on the force with which the object is thrown". These two new answers involve a misconception about gravity: *Constant gravity does not imply constant acceleration* (Liu and Fang 2016) and the already cited *impetus dissipation* (Hestenes et al. 1992).

Fig. 6 Question 3 of our questionnaire. In bold the right answer

3. A body moves in uniform circular motion along a circumference with a radius of 1 m at a speed of 5 m/s. Which statement is correct?

a) Since it is uniform motion, the speed is constant and the acceleration is zero.
b) **The tangential acceleration is zero, and the centripetal acceleration is 25 m/s^2**
c) The tangential acceleration is zero, and the centripetal acceleration is 4 m/s^2
d) None of the above is correct.

Fig. 7 Question 7 of our questionnaire. In bold the right answer

7. If you drop an object without giving it any push, it falls with an acceleration of 9.8 m/s^2, neglecting air resistance. However, if you throw it downward with force, what will its acceleration be after it leaves your hand, still assuming there is no air resistance?

a) It is less than 9.8 m/s^2
b) It is greater than 9.8 m/s^2
c) It is equal to 9.8 m/s^2
d) It depends on the initial velocity of the object

6 Conclusion

As seen in Sect. 4, the Classical Test Theory confirms that our questionnaire was well-structured and reliable. Concerning Table 4 we can observe that in the first 5 questions and in the last one students chose a distractor more often than the right answer and this could be a hint of the presence of specific misconceptions related to the topics covered by those questions. In addition, we can notice that only 3 "wrong answers" were chosen by fewer than 5% of the students so most of our distractors were effective. This allowed us to repeat this experience during the academic year 2023/2024 with the corrections exposed in Sect. 5.

In addition, in 2023/2024 we transformed the questionnaire into a 2-way test adding a Likert-scale question after each item. In this new question, we asked "how much do you feel confident about your answer?" with 4 possible choices. We think that this will help us better recognize misconceptions of students.

We also started to estimate the effectiveness of our course by observing the results of our students in the actual TOL and the first results seem to be interesting. In addition, we are analyzing the results in the updated propaedeutic questionnaire performing a new validation.

References

Alhadabi A (2021) Front Psychol 12:634120

Bandalos DL (2018) Measurement theory and applications for the social sciences. Guilford Publications

Bandura A (1997) Self-efficacy: the exercise of control. W.H. Freeman and Company, New York

Corsi di orientamento attivo nella transizione scuola università (2023). https://www.polimi.it/orientamento-pnrr

Crocker L, Algina J (1986) Introduction to classical and modern test theory. ERIC

Decreto ministeriale 934/2022 (2022)

Ding L, Beichner R (2009) Phys Rev Spec Top-Phys Educ Res 5(2):020103

Doran RL (1980) Basic measurement and evaluation of science instruction. ERIC

Ebel RL, Frisbie DA (1972) Essentials of educational measurement. Prentice-Hall Englewood Cliffs, NJ

Hattie J, Clarke S (2019) Visible learning: feedback. Routledge

Hazari Z, Potvin G, Tai RH, Almarode J (2010) Phys Rev Spec Top-Phys Educ Res 6(1):010107

Hestenes D, Wells M, Swackhamer G et al (1992) Phys Teach 30(3):141

Hu X, Jiang Y, Bi H (2022) Int J STEM Educ 9(1):47
Kline P (1986) A handbook of test construction: introduction to psychometric design. Methuen
Liu G, Fang N (2016) Int J Eng Educ 32(1):19
Malespina A, Singh C (2024) Eur J Phys
Maloney DP, O'Kuma TL, Hieggelke CJ, Van Heuvelen A (2001) Am J Phys 69(S1):S12
Romainville M (1994) Stud High Educ 19(3):359
Socrative. https://www.socrative.com/
The R project for statistical computing. https://www.r-project.org/
Trumper R (2003) Sci Educ 12:645
What the test consists of and how to prepare (2024). https://www.polimi.it/en/futuri-studenti/admission-to-laurea-programmes/engineering/what-the-test-consists-of-and-how-to-prepare
Wilcox BR, Lewandowski H (2016) Phys Rev Phys Educ Res 12(2):020132

Teacher Education

Physics Teaching in Primary School: Perception of Importance and Preparation of In-Service and Future Teachers

Tommaso Matteuzzi, Franco Bagnoli, Dario Giannerini, Francesca Nerattini, Giovanna Pacini, and Samuele Straulino

Abstract This study investigates the prominence of physics topics in the first cycle of education within the Italian school system and examines whether physics receives less emphasis compared to other scientific disciplines. Additionally, it explores the correlation between the importance teachers attribute to various topics and their perceived preparedness to teach these topics. To achieve these objectives, we devised two questionnaires and administered them to a sample of approximately 600 individuals, representative of the Italian school teacher population. Through this analysis, we aim to provide insights into the status of physics education within the Italian school system and identify potential areas for improvement in teacher training and curriculum development.

1 Introduction

In a world increasingly influenced by rapid technological advancements, the push for STEM education has brought increased emphasis on integrating physics into primary school curricula (Freeman et al. 2014). Early exposure to STEM subjects is believed to foster sustained interest and competence in these areas, which are essential for preparing students to a technologically advancing world. Notably, approaches that

T. Matteuzzi (✉) · F. Bagnoli · D. Giannerini · F. Nerattini · G. Pacini · S. Straulino
Department of Physics and Astronomy and CSDC, University of Florence, Sesto Fiorentino, Italy
e-mail: tommaso.matteuzzi@unifi.it

F. Bagnoli
e-mail: franco.bagnoli@unifi.it

G. Pacini
e-mail: giovanna.pacini@unifi.it

S. Straulino
e-mail: samuele.straulino@unifi.it

I. Testa et al. (eds.), *Connecting Physics Education Research and Practice*,
Challenges in Physics Education, https://doi.org/10.1007/978-3-031-86609-8_14

incorporate hands-on, experimental learning are shown to improve both engagement and comprehension (Maltese and Tai 2011; Dorouka and Kalogiannakis 2023).

Despite this, achieving positive attitudes toward physics among primary school teachers remains difficult. Many educators perceive physics as more abstract, complex, and secondary compared to other sciences such as biology and Earth sciences (Pahl and Tschiesner 2022). This perception is not isolated to one region but is observed in different countries (Margot and Kettler 2019). Research indicates that teachers often feel under-prepared to teach physics effectively, largely due to training that emphasizes theoretical knowledge over practical, experimental methods, which are crucial for primary education (Osborne et al. 2003; Vlachos et al. 2024; Banilower et al. 2012; Ingersoll et al. 2014).

Our study is mainly concerned with the Italian context. Here, physics holds a solid place in the national curriculum for primary education as established by the Italian Ministry of Education's guidelines (Ministero dell'Istruzione dell'Universitá e della Ricerca 2012). However, many primary school teachers share the perception of physics as challenging and secondary to other sciences. In response to this, Italian university programs for primary education have recently increased their focus on physics (Immè et al. 2022; Michelini and Stefanel 2022; Michelini et al. 2015). Despite these efforts, teachers often report a lack of confidence in teaching physics, especially when it involves experimental applications and real-world connections (Chiofalo et al. 2023; Amabile et al. 2022).

One hypothesis is that this issue stems from an educational background that, at the secondary level, places a strong emphasis on the theoretical and mathematical aspects of physics, often neglecting everyday applications and the experimental approach critical for primary education (Pahl and Tschiesner 2022). This theoretical focus likely affects how teachers prioritize and perceive physics topics relative to other sciences (Guarino et al. 2018).

Our aim was to carry out a preliminary study, exploratory in nature, to investigate which physics topics are actually covered in Italian primary schools and examine the relative importance teachers' assign to them. Moreover, we would evaluate how prepared in physics primary school teachers and Sciences of Primary Education (SPE) students feel, relative to other scientific subjects. Specifically, we addressed the following research questions:

(1) What physics topics are actually covered in Italian primary schools?
(2) How important are physics topics to teachers compared to other scientific disciplines?
(3) How prepared do primary school teachers and SPE students feel in scientific subjects, do they feel less prepared in physics?

Answering these question for the Italian school can reveal potential biases or priorities that may shape curriculum and influence student interest. Additionally, we were interested to explore any observable differences in attitudes toward scientific disciplines between primary school teachers and the general population. Understanding these differences can guide professional development and curriculum design, helping ensure teachers feel prepared to foster scientific curiosity.

Our study considers three participant groups: in-service primary school teachers, SPE students (i.e., pre-service teachers), and the general population with an interest in physics. To explore these questions, we developed two questionnaires: one for in-service teachers and SPE students, and another for the general population with an interest in physics. This comparative approach aims to determine whether teachers' views on the importance of physics align with those of the broader public. Even though our findings are specific to the Italian context, they can serve as guide in other countries where literature suggests the presence of similar issues.

Finally, based on these results, our research included a field study to assess primary school pupils' ability to connect theoretical knowledge in areas such as motion, electricity, and magnetism to practical problem-solving skills. This part of the study will be further explored in a subsequent publication.

2 Materials and Methods

Several well-established questionnaires exist to assess science teaching self-efficacy, an example is the STEBI (Riggs and Enochs 1990). However, to the best of our knowledge, there is no established tool designed to measure teachers' attitudes toward specific topics in physics and to compare these attitudes to those in other scientific disciplines. Given this gap, we developed a tailored, yet simple, questionnaire to directly address our research questions.

2.1 Teachers and SPE Student Questionnaire

The teachers' questionnaire is designed to investigate two primary aspects: first, the importance attributed to physics topics compared to other scientific disciplines and, second, whether any "lower importance" of physics topics may correlate with a lower level of perceived preparation in the subject.

To achieve this, we focused on three dimensions (De Rose 2018; Diamond et al. 2015): (1) awareness of the subject each topic belongs to (e.g., magnets as physics, plants as biology), (2) the perceived importance to teach each topic, and (3) self-evaluation of teachers' preparation in each subject (both rated on a scale from 1 to 10).

The questionnaire (included in the appendix) was organized into four sections with a total of 11 items. Table 1 provides an overview of the thematic dimensions and indicators used. Additionally, demographic and professional variables, such as age, gender, highest educational qualification, and two items related to the participants' current working position (covered in Sect. 3 of the questionnaire), were also collected. While this demographic data is not explicitly analyzed in the current study, it was gathered to gain a more comprehensive understanding of the sample's composition and to inform the design of future research where these demographic variables may be examined in greater detail.

Table 1 Questionnaire for SPE students and teachers. Hypotheses, indicators, item

Hypothesis	In primary school, topics recognized as physics topics are taught less frequently than those in other scientific disciplines			
Dimensions	Teacher preparation	Awareness of individual topic subjects		Teaching of individual topics
Indicators	Degree of preparation	Teaching timeline	Degree of importance	Subject attribution
Items	4.1 How prepared do you feel to teach the following topics?	4.2 When would you predominantly teach each of the following topics to students?	4.3 How important do you consider each of the following topics in primary school?	4.4 To which subject do you predominantly assign each topic?
Options	– Not prepared – Fairly prepared – Well prepared	Cardinal numbers from 1st to 5th or *not necessary* (to teach it)	Score 1–10	– Earth sciences, – Chemistry – Biology – Physics – Other

2.2 General Population Questionnaire

The questionnaire for the general population was designed to examine whether teachers' views on the importance of covering physics topics align with broader public opinion.

To explore the level of importance that the Italian population (beyond primary school teachers) attributes to the scientific topics in question, a selection of items from the teacher questionnaire was made, focusing specifically on the "degree of importance" indicator. The initial section of the general population questionnaire mirrors that of the teacher questionnaire.

2.3 Selection of Topics to be Investigated

The choice of topics in our survey is based on the TIMSS 2007 questionnaire (which is available on line, for example, on the INVALSI website in Italian: questionnaire for primary school teachers, fourth class, IEA, 2007 (Mullis et al. 2008)). We structured our questionnaire around TIMSS topics since it is a tested and effective model, internationally recognized. We used the 2007 TIMSS since this cycle marks the highest recorded performance of Italian students in TIMSS history.

In order to investigate the prevalence of teaching of some scientific disciplines compared to others, the questionnaires included topics about biology, physics, and Earth sciences. To have a more precise overview of the physics topics, the structure of the TIMSS questionnaire was modified, dedicating approximately 50% of the space to physics and leaving approximately 25% to each of the other two subjects (the comparison between the topics of the TIMSS questionnaires 2007 and those we examined and reported in Supplementary Table 5).

2.4 Selection of Samples and Administration of Questionnaires

All subjects were selected by disseminating the questionnaire via social networks (Facebook). The sample of teachers and SPE students was selected from groups made up of at least 200 people, who had one or more of the following keywords in the group name: *insegnanti/docenti/maestre/maestri* (teachers), *formazione, scuola, primaria, elementari* (formation and primary school), *scienze* (sciences), *fisica* (physics), *chimica* (chemistry), and *biologia* (biology).

Additionally, two mailing lists were created to present the project and request questionnaire participation. One mailing list targeted Italian comprehensive institutes (primary school teachers), and the other targeted teachers holding the chair of physics in "Primary Education Sciences" courses at Italian universities.

The sample size of the teaching population is 546, while that of students is 220.

The size of the target population of primary school teachers (in Italy about 298000) was found on the website of the Italian Ministry of Education and Merit. The student population (about 37000) was estimated based on number of available access to the Italian Primary Education Sciences courses in the last 5 academic years.

The population to which the generic questionnaire was submitted was selected from groups composed of at least 200 people and which had one of these words or a combination of the following keywords in the group name: *ricerca* (research), *fisica* (physics), *astronomia* (astronomy), *scienza/scienze* (sciences), and *matematica* (mathematics). The presentation of the project and the request to complete the questionnaire have been posted on these targeted Facebook groups.

2.5 Samples' Size and Representativeness

For each of the three populations, we collected as many responses as possible. A rough estimate of the required sample size can be calculated assuming a continuous random variable, a confidence level of 95% ($z = 1.96$), an effect size of 0.5-point difference ($e = 0.5$), and response variability of 5 points. Based on these parameters, the required sample size is estimated as $n_0 = z^2\sigma^2/e^2 \approx 384$. Thus, the teacher sample size allows us to achieve a precision of at least 95%. For the student sample, we estimate a precision of approximately 80% with a 5% margin of error (Diamond et al. 2015; Corbetta 2015).

The sample size obtained for the general population is 649 units. However, the total size of this target population is unknown and it is not possible to estimate a confidence level (Diamond et al. 2015).

Table 2 shows the values of the population and of the sample of teachers, divided by gender and age (source: Ministry of Education and Merit). The sample of teachers is representative of the target population, with a maximum discrepancy of less than 2% in the comparison of the percentages of both the individual variables and in the comparison of the percentages among multiple variables.

Table 2 Profile of Italian primary education teachers' population and sample (Semi-integers due to adjusted age group split)

	Up to 34	35–44	45–54	Over 54
Total population	26240	69862	105016	97340
Total sample	52.5	128.0	194.5	171.0
% population	8.8%	23.4%	35.2%	32.6%
% sample	9.6%	23.4%	35.6%	31.3%
Total female population	24455	66449	100566	94139
Female sample	50.0	123.0	188.0	166.0
% female population	8.6%	23.3%	35.2%	33.0%
% female sample	9.5%	23.3%	35.7%	31.5%
Total male population	1785	3413	4450	3201
Male sample	2.5	5.0	6.5	5.0
% male population	13.9%	26.6%	34.6%	24.9%
% male sample	13.2%	26.3%	34.2%	26.3%

For the SPE population, it was not possible to retrieve data on the entire population, as only certain universities make this information available. Regarding the general population, we focused specifically on individuals with an interest in physics-related topics, making it challenging to define a precise reference population. For these reasons, it was not possible to trace the characteristics of the population and therefore there are no data to verify their representativeness. However, the results obtained from the student and general samples will be used only as a comparison of the trend with respect to the teacher sample, without relying on them for statistical findings.

Finally, a comparison was made among the distributions of multiple variables, identifying the erroneous data. In particular, the distributions of items 4.2 and 4.3 were compared, as there is a close correlation between the importance of teaching a specific topic and its distribution over the primary years. The questionnaires affected by large discrepancies (teacher sample = 142, student sample = 15, general sample = 34) have been discarded.

3 Results

3.1 Attribution of Topics to Physics

To identify the topics of interest for subsequent analyses, i.e., those that the teachers themselves identify as physics topics, we started by analyzing the responses of the sample of teachers to item 4.4 of the questionnaire: "To which subject do you mainly attribute each individual topic?".

The results of the analysis are reported in Fig. 1. We note that most of the topics are assigned to the same discipline by a high percentage of teachers. Some topics,

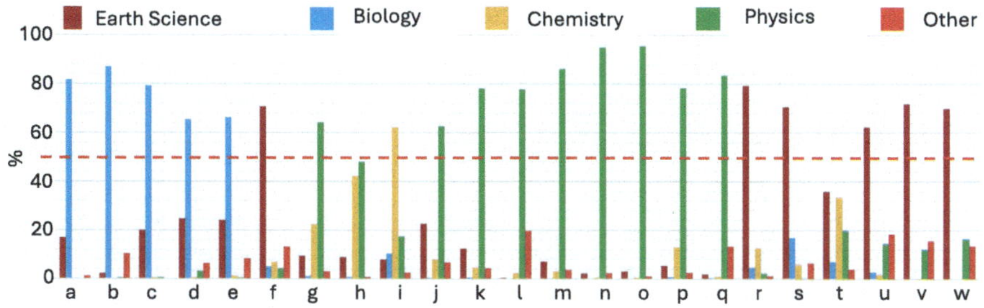

Fig. 1 Frequency of attribution of topics to subjects (Item 4.4). Letters correspond to topics as in Table 3

Table 3 Legend of considered topics. In the following plots, each letter corresponds to a specific topic as outlined in the table below

(a) Types of living beings and their characteristics	(f) Environmental changes	(k) Sources of light and related phenomena	(p) Heat flux and temperature	(u) Meteorological conditions
(b) Parts of the body and their functions	(g) Classification of materials and objects	(l) Electrical circuits	(q) Production of waves and sound	(v) Solar system
(c) Reproduction of plants and animals	(h) States of matter	(m) Property of magnets	(r) Rocks, minerals, sands	(w) Rotation of earth
(d) Behavioral models	(i) Common transformation of materials	(n) Motion of objects, velocity	(s) Fossils of plants and animals	
(e) Ecological relations	(j) Common forms of energy and their usage	(o) Types of forces (e.g., gravity) and their effects	(t) Air	

however, have a percentage corresponding to the mode lower than or equal to two-third of the sample (for example, g = classification of objects/materials based on characteristic properties, h = states of matter).

3.2 *Importance Attributed to Different Scientific Subjects*

The topics on which we concentrated our analyses are those attributed to physics by the majority of the sample. We analyzed their perceived importance in item 4.3, where teachers were asked to assign a degree on a scale of 1–10 to each of them. In Fig. 2, the average importance values for each topic are reported for the teacher and the student samples. A substantially identical trend is observed. We assessed the relative importance of each topic by comparing its level of importance against all other topics

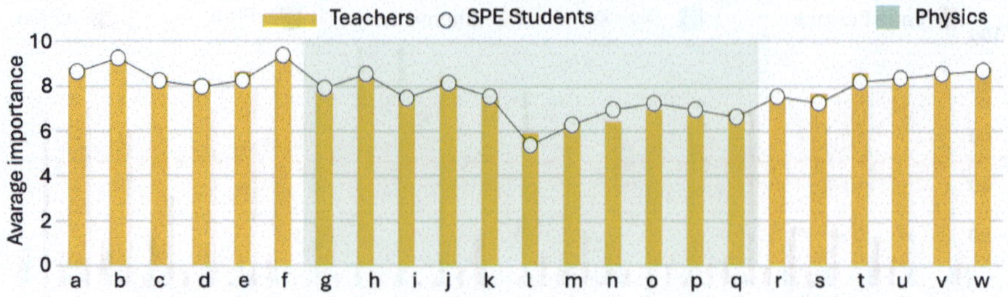

Fig. 2 Average score of the importance attributed to topics (as in Table 3) by teachers and SPE students (Item 4.3)

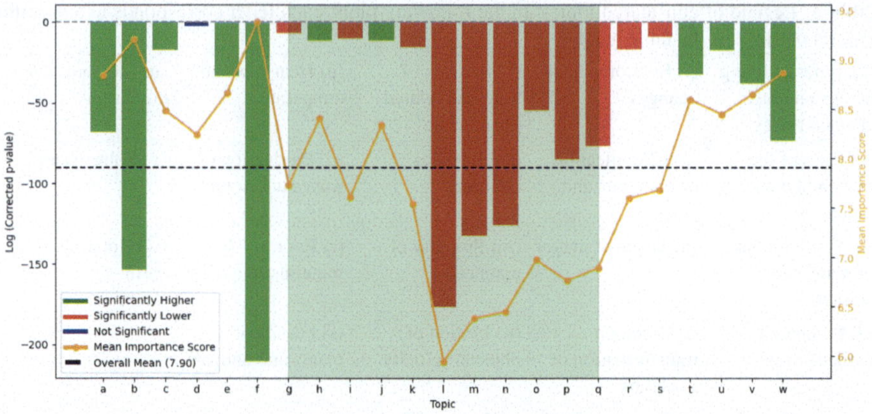

Fig. 3 Importance attributed to topics. A non-parametric Mann–Whitney U test was used to assess the significance of different levels of importance. Green (red) bars represent topics considered more (less) important by the teachers. The yellow line gives the mean score of importance. Physics topics are highlighted with a light green background

using the non-parametric Mann–Whitney U test. This test evaluates whether the distribution of importance scores for a specific topic is significantly higher (or lower) compared to the distribution of scores for other topics. The results are summarized in Fig. 3, which highlights topics with statistically significant differences in perceived importance.

Also analyzing the percentages of negative reviews confirms a difference between the subjects examined, Fig. 4. For biology we find a range of values between 4 and 7%, for Earth sciences between 3 and 13%, while for physics a minimum value of 6% and a maximum value of 36%. This means that topics identified as "physics" by the sample of teachers obtained a greater frequency of negative opinions compared to the other two subjects, with a difference of 10–16% on the maximum values.

We can therefore say that the sample of teachers considers "less important" to address physics topics.

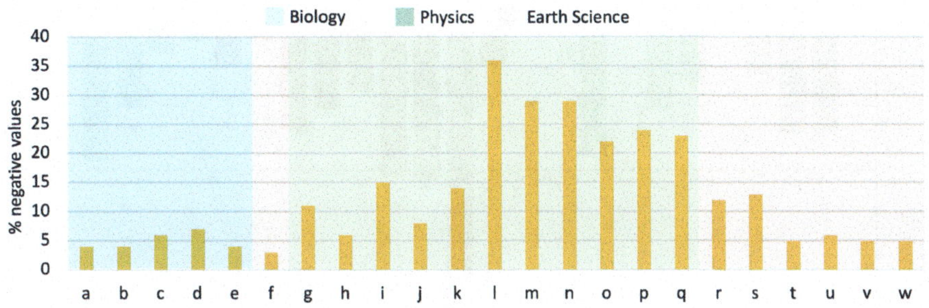

Fig. 4 Negative value attributed to topics (as in Table 3)

3.3 *Self-perception of the Preparation of Teachers and SPE Students*

The detection of teacher preparation aims at testing whether the importance of the topics identified is related to the confidence with them.

In item 4.1 of the questionnaire "How much prepared do you feel to teach the following topics?" one is asked to choose between: "Not prepared", "Fairly prepared", and "Very prepared".

Figure 5a shows the response frequencies for the different topics. We note how the percentages of "Not prepared" are higher for physics topics. The substantial

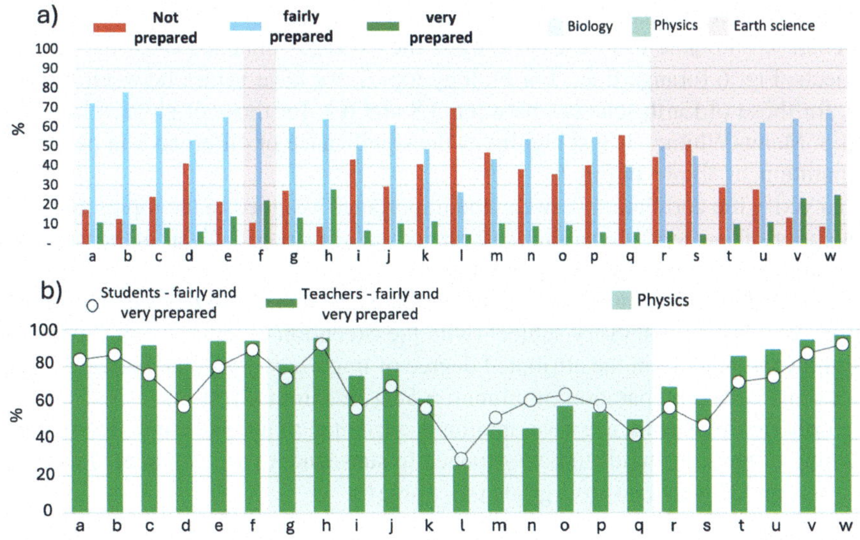

Fig. 5 **a** Average value of preparation estimate of teachers. **b** Comparison between the average values of preparation estimate of teachers and students

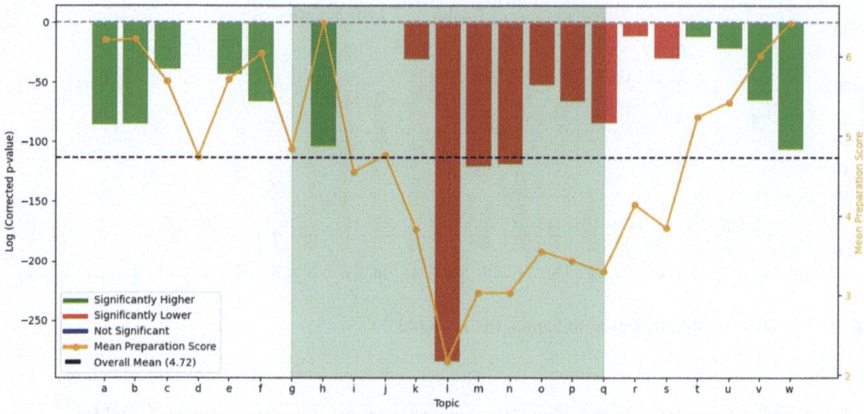

Fig. 6 Teachers perceived preparation on topics. As before, non-parametric Mann–Whitney U test was used to assess the significance of different levels of preparation. Green (red) bars represent topics for which teachers feel more (less) confident. The orange line indicates the mean score of confidence. Physics topics are highlighted with a light green background

difference in the average results between biology and Earth science topics with respect to physics topics highlights a level of preparation perceived by primary school teachers for this last subject between "Not prepared" and "Fairly prepared", lower than for the other subjects (Fig. 5b).

To build a single indicator for preparation, we assigned a value of 1 to the answer "Not prepared", a value of 5 to the answer "Fairly prepared", and a value of 10 to the answer "Very prepared".

There are notable differences between the averages obtained from the various subjects, Fig. 6 (orange line). For biology topics, we have values between 4.7 and 6.4, for those of Earth sciences between 3.8 and 6.4, for those of physics from 2.1 to 4.8. Mann–Whitney U test highlights the significance of differences in perceived preparation, as shown in Fig. 6.

To verify the correlation between the importance assigned to the topics and self-perception of teachers' preparation, we carried out a bivariate analysis of the variables "preparation" and "importance of the topic".

Table 4 shows the change in positive judgment in relation to the reported preparation. We observe that every subject (with the exception of topics a and f) presents a significant increase in the attributed degree of importance with the increase in the level of perceived preparation (minimum = 58%, maximum 100%).

Furthermore, when a negative opinion is attributed to a topic, we notice that the teaching of the same tends to be postponed to subsequent years, compared to those who responded with a positive grade (see Table 6).

Table 4 A bivariate analysis was conducted to examine the correlation between teachers' perceived preparation and the importance they assign to each topic. For each topic, we report the proportion of teachers who feel, respectively, unprepared, fairly prepared, or very prepared, and assign high importance to that topic. The subject areas attributed to each topic are indicated by: red for Earth sciences, blue for biology, green for physics, and white not assigned

Topic	% Pos./not prepared	% Pos./fairly prepared	% Pos./very prepared
(a) Types of living beings and their characteristics	100	94	99
(b) Parts of the body and their functions	86	97	97
(c) Reproduction of plants and animals	92	94	96
(d) Behavioral models	89	93	98
(e) Ecological relations	89	96	98
(f) Environmental changes	100	96	98
(g) Classification of materials and objects	79	91	93
(h) States of matter	89	93	97
(i) Common transformation of materials	75	87	91
(j) Common forms of energy and their usage	88	93	94
(k) Sources of light and related phenomena	78	91	93
(l) Electrical circuits	58	78	100
(m) Property of magnets	63	78	96
(n) Motion of objects, velocity	66	76	87
(o) Types of forces (e.g., gravity) and their effects	70	83	93
(p) Heat flux and temperature	65	85	93
(q) Production of waves and sound	68	85	94
(r) Rocks, minerals, sands	77	91	98
(s) Fossils of plants and animals	76	93	95
(t) Air (composition, etc.)	89	95	99
(u) Meteorological conditions	89	94	98
(v) Solar system	91	95	97
(w) Rotation of Earth	79	94	93

3.4 Comparison Between Teachers and the General Population

For the generic sample (i.e., individuals who do not work as primary school teachers, are not PSE students but have an interest in physics topics), the only question in the questionnaire was defined by item 4.3: "How important do you consider the following topics in primary school?"

Figure 7 shows that even the generic population considers physics topics less important to be addressed in primary school compared to topics in other subjects. It can be observed that the absolute minimum is related to the topic of electrical circuits, followed by properties of magnets.

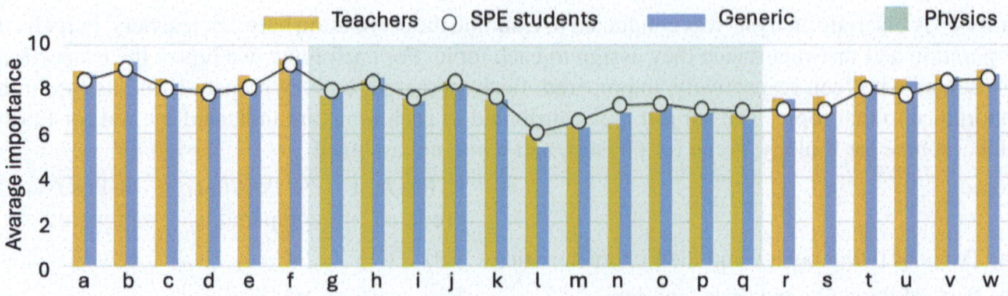

Fig. 7 Average value of importance for teachers, SPE students, and general population sample

4 Conclusions

In this study, we aimed to explore attitudes toward physics education among primary school teachers, students in primary education sciences (SPE student, future teachers), and the general population with an interest in physics. Our research questions focused, primarily, on understanding the perceived importance of physics topics in primary school, comparing attitudes toward different topics and examining how self-assessed preparation levels impact these perceptions.

Our findings reveal a consistent undervaluation of physics topics compared to biology and Earth science in the primary school curriculum. This trend is observed not only among teachers but also among non-teachers who have an interest in physics. A key result is the correlation between a self-perceived lack of preparation in physics among current and future teachers and the lower importance attributed to these topics, suggesting that confidence in teaching physics may directly influence its prioritization in the classroom.

These insights highlight a potential gap in teacher preparation programs and suggest the need for targeted training in physics education to ensure teachers feel confident in these areas. We identify specific physics topics, such as motion, electricity, and magnetism, as particularly challenging, even for educators.

The study's scope was limited to a representative sample of teachers within Italy, but results are in line with findings in other countries (Pahl and Tschiesner 2022; Margot and Kettler 2019). Future research could build on this work by examining interventions that improve physics preparation among primary school teachers and assessing the long-term impact of these interventions on student engagement with physics.

Appendix: Questionnaires

Common part

1. Personal data and training

    1.1 Age: options (under 25, 25–29, 30–39, 40–49, 50–59, 60 or older);
    1.2 Sex;
    1.3 Educational qualification;

1.4 Specify diploma-degree-master's/doctorate.

2. Authorization to process personal data

2.1 Processing of personal data.

Teachers

3. Work

3.1 Do you teach in primary school and/or are you a Primary Education Science student?: options (yes, no);
3.2 Occupation: options (student, student/worker, worker).

4. Attitude toward scientific subjects

4.1 How prepared do you feel to teach the following topics? options (not prepared, fairly prepared, very prepared);
4.2 When would you predominantly teach each of the following topics to your students? Mark the year in which you consider it appropriate to address, or begin to address, a specific topic: options (1st, 2nd, 3rd, 4th, 5th, I don't think it necessary to teach it in primary school);
4.3 How important do you think the following topics are in primary school? Give a score from 1 to 10, where 1 equals "not at all important" and 10 equals "extremely important": options (numbers from 1 to 10);
4.4 To which subject do you mainly attribute each individual topic? Since many topics listed are interdisciplinary, we ask you to identify for each topic the subject that mainly addresses it: options (Earth science, biology, chemistry, physics, other).

Generic

5. Work

5.1 Do you teach in primary school and/or are you a Primary Education Science student?: options (yes, no);
5.2 Occupation;
5.3 Specify job.

6. Attitude toward scientific subjects

6.1 How important do you think the following topics are in primary school? Give a score from 1 to 10, where 1 equals "not at all important" and 10 equals "extremely important": options (numbers from 1 to 10).

Supplementary Tables
See Tables 5 and 6.

Table 5 Comparison between TIMMS topics and our questionnaire

Biology	
…	**a) Types of living beings, their characteristics and classification**
- Main structures of the body and their function in humans and other organisms *(plants and animals)*	**b) Main structures of the body and their function in humans**
- Reproduction and development of plants and animals *(transmission of general characteristics, life cycles of common organisms)*	**c) Reproduction and development of plants and animals (transmission of general characteristics, life cycles of common organisms)**
- Physical characteristics, behavioural patterns and survival of organisms living in different environments	**d) Physical characteristics, behavioural patterns and survival of organisms living in different environments**
- Relationships in a community of living things *(e.g., simple food chains, predator/prey relationships)*	**e) Relationships in a community of living beings *(e.g., simple food chains, predator/prey relationships)***
- Environmental changes *(effects of human activities, pollution, and prevention)*	**(f) Environmental change *(effects of human activities, pollution, and prevention)***
- health *(e.g., transmission/prevention of contagious diseases, disease health and symptoms, diet, exercise)*	…
Physics	
- Classification of objects/materials according to physical properties *(e.g., mass, shape, volume, colour, hardness, composition, electrical and thermal conductivity, magnetic attraction)*	**g) Classification of objects/materials according to their characteristic properties *(e.g., mass, shape, volume, hardness, composition)***
- Formation and physical separation of mixtures	…
- States of matter *(solid, liquid, gas)* **and differences in their physical properties** *(shape, volume)* **including changes in the state of matter due to heating and cooling** *(melting, freezing, boiling, evaporating, and condensing)*	**(h) States of matter *(solid, liquid, aeriform)* and differences in their properties *(shape, volume)* including changes in the state of matter due to heating and cooling *(melting, freezing, boiling, evaporating, and condensing)***
-Common transformations in materials *(e.g., decomposition of animal/plant material, combustion, rusting, cooking)*	**i) Common transformations in materials (e.g., decomposition of animal/plant material, combustion, rusting, firing)**
- Common energy sources/forms and their practical uses *(e.g., wind, sun, electricity, fuel combustion, water wheel, food)*	**j) Common energy sources/forms and their practical uses** *(e.g., wind, solar, fuel combustion, water wheel)*
- Light *(e.g., sources and behaviour)*	**k) Common light sources and related phenomena** *(e.g., rainbow and shadow formation, object visibility, mirrors, colours)*
- Electrical circuits	**l) Electrical circuits**
- Properties of magnets	m) Properties of magnets (e.g., north, and south poles, magnetic attraction, and repulsion)
…	**n) The movement of objects, the concept of speed**
- Forces that cause objects to move *(e.g., gravity, pushing/pulling)*	o) Types of forces (including gravity) and their effects on the motion of objects
…	p) Heat Flow and Temperature
…	q) Production of sound by vibration
Earth Sciences	
- Morphological features of the Earth *(e.g., mountains, plains, rivers, deserts)*	…
…	**r) Rocks, minerals, sand, and soil**
- Fossils of animals and plants *(age, formation)*	s) Fossils of animals and plants (age, formation)
- Water on Earth *(location, types, and movement)*	…
- Air *(composition, proof of its existence, uses, and importance for life)*	**t) Air *(composition, proof of its existence, uses and importance for life)***
- Relationships between the morphological characteristics of the Earth *(e.g., mountains, plains, rivers, deserts)* **and human use** *(e.g., agriculture, irrigation, land development)*	…
- Weather conditions day after day and throughout the seasons	**u) Weather conditions day after day and over the course of the seasons**
- Earth's solar system *(planets, Sun, Moon)*	**v) The solar system** *(planets, Sun, Moon)*
…	w) Rotation of the Earth (e.g., day and night)

(red) deleted arguments, (green) topics kept in full, (viola) partially modified arguments.

Table 6 Correlation of attribution of importance and teaching year. Colors represent subject areas: red = Earth sciences, blue = biology, green = physics, and white = not assigned

Topics	Negative review		Positive review	
	Mode	% not nec./ negative	Mode	% not nec./ positive
(a) Types of living beings and their characteristics	3°	4	3°	0
(b) Parts of the body and their functions	5°	5	5°	0
(c) Reproduction of plants and animals	4°	6	3°	1
(d) Behavioral models	3°	13	4°	1
(e) Ecological relations	3°	0	3°	0
(f) Environmental changes	1°	0	4°	0
(g) Classification of materials and objects	5°	13	3°	1
(h) States of matter	3°	6	3°	0
(i) Common transformation of materials	Not necessary	33	4°	3
(j) Common forms of energy and their usage	5°	19	5°	2
(k) Sources of light and related phenomena	5°	37	5°	3
(l) Electrical circuits	Not necessary	79	5°	12
(m) Property of magnets	Not necessary	59	5°	8
(n) Motion of objects, velocity	Not necessary	70	5°	12
(o) Types of forces (e.g., gravity) and their effects		57	5°	9
(p) Heat flux and temperature	Not necessary	55	5°	10
(q) Production of waves and sound		50	5°	6
(r) Rocks, Minerals, sands	Not necessary	31	3°	1
(s) Fossils of plants and animals	3°	34	3°	1
(t) Air (composition, etc.)	3°	20	3°	1
(u) Meteorological conditions	3°	29	2°	2
(v) Solar system	5°	4	5°	2
(w) Rotation of Earth	Not necessary	23	5°	1

References

Amabile A, Annunziata A, Artiano G, Balzano E (2022) Experimentation and research in the physics course for the preparation of primary school teachers in Naples. Educ Sci 12(4):241

Banilower ER, Smith PS, Weiss IR, Malzahn KA, Campbell KM, Weis AM (2013) Report of the 2012 national survey of science and mathematics education. Horizon Research Inc, Chapel Hill, NC

Chiofalo ML, Corradini O, De Angelis I, Falomo Bernarduzzi L, Malgieri M, Sutrini C, Giliberti M, Immè J, Michelini M, Organtini G, Pagliara S, Pavesi M, Sabbarese C, Salamida F, Straulino S (2023) Physics for primary school teachers in Italy: comparative analysis in a dedicated survey. In: Physics teacher education. Springer Nature Switzerland, pp 163–171

Corbetta P (2015) La ricerca sociale: metodologia e tecniche, Il Mulino (2015)

De Rose C (2018) L'indagine campionaria e il sondaggio d'opinione. Metodi quantitativi della ricerca sociale, Carocci Editore

Diamond I, Foster L, Jefferies J (2015) Beginning statistics: an introduction for social scientists, 2nd edn. Sage Publications Ltd

Dorouka P, Kalogiannakis M (2023) Teaching nanotechnology concepts in early-primary education: an experimental study using digital games. Int J Sci Educ. https://doi.org/10.1080/09500693.2023.2286299

Freeman B, Marginson S, Tytler R (eds) (2014) The age of STEM: educational policy and practice across the world in Science, Technology, engineering and mathematics, 1st edn. Routledge. https://doi.org/10.4324/9781315767512

Guarino A, Marvilliers S, Pacini G, Bagnoli F (2018) Improving science teaching in primary schools: FoCoSTEP, a French experience of training support. In: Formazione & Insegnamento, pp 241–250

Immè J, Michelini M, Chifalo ML, De angelis I, Falomo Bernarduzzi L, Fazio C, Giliberti M, Pagliara S, Pavesi M, Sabbarese C (2022) Le questioni affrontate durante il convegno "Quale didattica della fisica per formare gli insegnanti di scuola primaria?"—Which kind of physics education for prospective primary teachers?. In: SUPPLEMENTO PLS-FISICA-FIP, GIORNALE DI FISICA DELLA SOCIETÁ ITALIANA DI FISICA

Ingersoll RM, Sirinides P, Dougherty P (2014) Seven trends: the transformation of the teaching force. Consortium for Policy Research in Education

Maltese AV, Tai RH (2011) Pipeline persistence: examining the association of educational experiences with earned degrees in STEM among U.S. students. Sci Educ 95:877–907. https://doi.org/10.1002/sce.20441

Margot KC, Kettler T (2019) Teachers' perception of STEM integration and education: a systematic literature review. Int J STEM Educ 6(2):2. https://doi.org/10.1186/s40594-018-0151-2

Michelini M, Santi L, Stefanel A (2015) La formazione degli insegnanti in fisica come sfida di ricerca: problematiche, modelli, pratiche. Italian J Educ Res 191–208

Michelini M, Stefanel A (2022) Prospective primary teachers and physics Pedagogical Content Knowledge's. In: Research Unit in Physics Education, Dept. of Chemistry, Physics and Environment. https://www.fisica.uniud.it/urdf/articoli/ftp/2011/2011-12.pdf

Ministero dell'Istruzione dell'Universitá e della Ricerca (2012) Indicazioni nazionali per il curriculo della scuola dell'infanzia e del primo ciclo di istruzione. https://www.miur.gov.it/documents/20182/51310/DM+254_2012.pdf

Mullis IVS, Martin MO, Foy P, Olson JF (2008) International association for the evaluation of educational achievement (IEA), TIMSS & PIRLS international study center, trends in international mathematics and science study: TIMSS 2007: international mathematics report: findings from IEA's trends in international mathematics and science study at the fourth and eighth grades. TIMSS & PIRLS International Study Center, Boston College, Chestnut Hill, MA

Osborne J, Simon S, Collins S (2003) Attitudes towards science: a review of the literature and its implications. Int J Sci Educ 25(9):1049–1079. https://doi.org/10.1080/0950069032000032199

Pahl A, Tschiesner R (2022) What influences attitudes and confidence in teaching physics and technology topics? An investigation in kindergarten and primary-school trainee teachers. Sustainability 14(1):87. https://doi.org/10.3390/su14010087

Riggs IM, Enochs LG (1990) Toward the development of an elementary teacher's science teaching efficacy belief instrument. Sci Educ 74(6):625–637. https://doi.org/10.1002/sce.3730740605

Vlachos I, Stylos G, Kotsis KT (2024) Primary school teachers' attitudes towards experimentation in physics teaching. Eur J Sci Math Educ 12(1):60–70. https://doi.org/10.30935/scimath/13830

Energy Content Knowledge of Primary School Teachers Engaged with Narratives of Forces of Nature

Chiara Puecher, Hans Ulrich Fuchs, Cliona Murphy, and Federico Corni

Abstract This chapter presents an instrumental collective case study, which explores the content knowledge related to energy of three primary school teachers after their engagement with narratives of Forces of Nature in a Continuing Professional Learning Programme. First, the rationale of the study and the conceptual framework of an Imaginative Approach to Physics in Primary Education, specifically its theoretical foundations grounded in cognitive linguistics, are presented. Secondly, the methodology with which the study was conducted is outlined, showing how teachers' expressions were analyzed applying conceptual metaphor theory. The discussion of the results shows how teachers developed an articulated and narrative conception of energy, emerging from an interweaving of metaphors.

1 Introduction

Energy is common to all science disciplines, and provides a unifying perspective to think about numerous phenomena (Millar 2005). Energy is a primary educational objective in science instruction and is a "cornerstone" for students' comprehension of physics' concepts at all levels (Toli and Kallery 2021). In Italy, for example energy is one of four main cognitive organizers with conceptual and cultural relevance in the first school cycle (Heron et al. 2009). Primary school teachers have to address the challenge to teach energy in ways that promote deep conceptual understanding.

C. Puecher (✉) · H. U. Fuchs · F. Corni
Free University of Bozen-Bolzano, Bressanone, Italy
e-mail: Chiara.Puecher@education.unibz.it

H. U. Fuchs
e-mail: HansUlrich.Fuchs@unibz.it

F. Corni
e-mail: federico.corni@unibz.it

C. Murphy
Dublin City University, Dublin, Ireland
e-mail: cliona.murphy@dcu.ie

I. Testa et al. (eds.), *Connecting Physics Education Research and Practice*,
Challenges in Physics Education, https://doi.org/10.1007/978-3-031-86609-8_15

Many studies investigate students' understanding of energy and their difficulties at all levels (Lee and Liu 2010; Papadouris and Constantinou 2016; Scherr et al. 2012; Trumper 1993). Moreover, much research deals with the role of teaching in helping students in this (Papadouris and Constantinou 2011; Delegkos and Koliopoulos 2020; Hertting 2016; Liu and McKeough 2005; Millar 2014; Neumann et al. 2013).

To support teachers in this challenge, an Imaginative Approach to Physics in Primary Education (IAPPE) and more specifically to Forces of Nature (FoNs) (Corni 2013; Corni and Fuchs 2020, 2021; Fuchs and Corni 2023) has been developed. The approach is based on theoretical frameworks in physics (Fuchs 2006, 2014), narratology (Bruner 1990; Herman 2009), cognitive linguistics (Hampe 2005; Lakoff and Johnson 1980, 1999), and on a developmental theory of cognitive tools (Egan 1997). It is grounded in the relationship between child and nature, promotes imaginative understanding of natural phenomena, building on the notion of Force of Nature (Corni and Fuchs 2020), and it is consistent with the imaginative nature of our mind (Egan 1997; Lakoff and Johnson 1980; Johnson 1987) and of the object of study, namely, physics (Fuchs 2006, 2014).

2 Conceptual Framework

2.1 Embodied Mind, Image Schema, and Conceptual Metaphor

The approach finds its roots in cognitive linguistics, specifically in studies concerning embodied mind, conceptual metaphor, and image schema. According to the embodied cognition perspective, our repeated sensory–motor interactions with the environment define how we are able to understand our experience and how consequently we act. In short, our mind is embodied (Johnson 1987; Lakoff and Johnson 1980, 1999). The concept of image schema emerged from the attempt to understand how meaning is structured from our bodily interactions with the environment (Johnson 1987; Lakoff 1987). Image schemas are "recurring patterns of our sensory-motor experience by means of which we can make sense of that experience and reason about it". Lakoff and Johnson (1980) claim the metaphorical nature of our conceptual system. Metaphor is then a cognitive tool that allows us to shed light on secondary, unfamiliar domains, which thus undergo an overall restructuring. In particular, a projection occurs from a known domain, referred to as the source domain, to a less familiar or abstract domain, the target domain (Lakoff and Johnson 1999). It is therefore concluded that our rationality is imaginative because the structures we use for reasoning and understanding are imaginative (Johnson 1987).

3 Imaginative Approach to Physics in Primary Education: The Notion of Force of Nature and the Energy Concept

3.1 The Notion of Force of Nature

A Force of Nature (FoN) is "a medium-scale perceptual unit or perceptual gestalt that is most easily associated with and recognized as any form of powerful agency" (Corni and Fuchs 2020; Fuchs 2006, 2015). Note that here the word "force" is not used as it is intended in mechanics. Forces are Gestalts (Fuchs 2006, 2007, 2010, 2014). In general, the gestalt of the FoN can be structured by means of the following image schemas: i. the "(fluidlike) substance" schema which gives metaphoric sense to the aspect of "quantity", "amount", or "size", and is linked to that of "flow", "container", and "carrier"; ii. the "(vertical) level" or "state" schema and the consequent ones of "difference of levels" and "states" (i.e., tension), which structure the understanding of the aspect of intensity, i.e., the (extended) potential; iii. the "direct manipulation" and "causation" schemas, which shape the aspect of power, connected to the concept of energy (Fuchs 2007, 2010). We abstract these schemas from our experience with phenomena, particularly with fluids, and, by analogy, use the same schemas to understand other phenomena such as heat, electricity, magnetism, substances, and motion (Corni and Fuchs 2020; Fuchs and Corni 2023).

3.2 Energy is Connected to the Aspect of Power and is a (Partly) Fluidlike Quantity

The concept of power, linked to that of intensity and quantity of a FoN, is the first step in understanding the concept of energy. In physics, energy is a quantity that indicates how much is happening in an interaction between two phenomena and how much of a phenomenon is caused by taking on the same roles in a causing phenomenon (Fuchs and Corni 2023). Power, on the other hand, is the rate at which energy is released or used in physical processes (Fuchs 1987, 1996; Maurer 1990). The concept of energy shares with that of FoN the schemas that characterize the (FLUIDLIKE) SUBSTANCE metaphor and are associated with it (Corni and Fuchs 2020). However, energy is not a fluidlike quantity in the sense of FoN, because it is made available by FoNs in their interactions (Fuchs and Corni 2023), it is a property of a Force. There is indeed a common ontology of energy as a substance-like quantity in physics research, in textbooks, and in the science education discourse (Amin 2009; Harrer 2017; Lancor 2014; Corni et al. 2019; Scherr et al. 2012). In Feynman (1963) energy is defined "a most abstract idea", "a mathematical principle", and lastly "anything concrete". However, according to several authors, conceiving energy as substance-like is not an obstacle to the development of a scientific understanding of it (Corni et al. 2019; Brewe 2011; Duit 1987; Millar 2005).

3.3 Energy Emerges in Relation to Aspects and Interactions of Forces of Nature

To address imaginatively the concepts of energy storage and transfer, the FoNs and specifically their fluidlike substance aspects are considered energy containers and carriers (Corni and Fuchs 2020). Respectively, the image schema "container" and "carrier" are metaphorically projected into the FoN experiential domain. Moreover, energy carriers flow from a system to another (Corni and Fuchs 2020). In this case, the reference image schema is "flow". Energy is carried by FoNs when they are tensed, at high potential. The transport of energy is thus connected also to the metaphors INTENSITY IS A VERTICAL LEVEL OR STATE and DIFFERENCE OF INTENSITY IS A LEVEL/STATE DIFFERENCE. Tension, the difference of two values of potential at two different locations in a physical system, is indeed primary: a tension is necessary, in particular, a tense extensive quantity, for something to happen (Fuchs and Corni 2023). Moreover, despite being present in the discourse of both laypeople and physicists (Corni et al. 2019), it is incorrect to speak of energy as a cause: energy "is not a description of a mechanism" (Feynman 1963). In interactions between FoNs, energy is made available by the agent FoN and accepted and used by the patient FoN. The FoN releasing the energy is thus metaphorically a powerful agent while the FoN receiving it is the patient (Corni and Fuchs 2020; Fuchs and Corni 2023). In addition, energy is neither produced nor destroyed; its quantity is conserved. Feynman (1963) discusses energy conservation by saying "it is just a strange fact that we can calculate some number and when we finish watching nature go through her tricks and calculate the number again, it is the same". Moreover, in traditional approaches to physics, energy is said to be dissipated. To understand this concept imaginatively, it is said that energy is partly missed by the patient FoN and used in the production of Heat (Fuchs et al. 2022; Fuchs and Corni 2023). As early as 1824, Sadi Carnot conceptualized Heat as something very similar to FoN, arguing that Heat has "a great motive power" and describing its effects. Lastly, energy is always the same. The concept of conversion or transformation of energy into different forms does not allow us to address the extensive quantities related to their potential and power (Corni and Fuchs 2021).

3.4 Forces of Nature Narratives

The theories and laws of physics have an imaginative and metaphorical nature; if we want to describe and model a concrete physical phenomenon, we appeal to something more elaborate than a sum of metaphors (Corni and Fuchs 2020). Scientists build models and develop simulations in which the dynamics of systems unfold in a temporal dimension. By analogy, simulations correspond to stories, i.e., concrete narratives, whereas models correspond to the "story-world" (Herman 2002), the background against which the stories are told (Corni and Fuchs 2020). Natural and technical phenomena therefore can be framed narratively (Fuchs 2015; Fuchs et al. 2018). The Gestalt of the FoN allows us to create and tell stories where these Forces emerge, act, and interact. Narrative can be experienced using various imaginative forms of expression, like oral expression or mimesis, namely, bodily expression. The stories of FoNs (Corni and Fuchs 2020) begin with FoNs tensions, develop as FoNs, characterized by their specific aspects, act, and interact. They finally end when the tensions are resolved, or a balance is established. In the FoNs Theater (Corni and Fuchs 2021), the narrative unfolds through the body and gestures of the actors, who represent the FoNs and interact exchanging energy.

4 Research Question

This study intends to apply the discussed aspects of an Imaginative Approach to Physics in Primary Education (PPE) within a Continuing Professional Learning Programme (CPLP) with primary school teachers and investigate, in particular, how the engagement with narratives of FoNs influenced teachers' energy content knowledge (Van Driel et al. 1998). The interest in linguistic expression is based on the assumption that the language we use reveals our conceptual system (Lakoff and Johnson 1980). The research question is the following: how do teachers linguistically express their energy-related content knowledge as a result of their engagement with narratives of Forces of Nature?

5 Methodology

The research method adopted was instrumental collective case study (Stake 1995). It differs from intrinsic case study because the ultimate interest is not the particular case. Studying the case is indeed instrumental in understanding something else, namely, the interest advanced by the research question. Moreover, to better address the research question, the study was extended to several cases, coordinated between each other, assuming the features of a collective case study.

6 Sample

Three primary school teachers from two different schools in an Italian province participated in the study. The following pseudonyms were given to the teachers: Anne, Barbara, Cameron. When the study took place, all the teachers were teaching science: Anne was teaching fourth grade (9–10 year-old students) and Barbara and Cameron fifth grade (10–11 year-old students). After secondary school, Anne focused her studies on a scientific area. She then taught in the primary school for more than

20 years, 18 of which she taught science. Barbara chose a humanistic area of study after secondary school. In total, she taught in primary school for more than 30 years out of which spent 10 teaching science. Cameron specialized in the Montessori method. She taught in the primary school for more than 10 years, of which she taught science for 1 year. Barbara, unlike the other teachers, was already familiar with the Imaginative Approach to PPE at the time of the study.

7 The Continuing Professional Learning Programme (CPLP): Teachers' Engagement with Narrative

The CPL Programme was divided into two phases. The first phase addressed the theoretical foundations and the resources of the approach, allowing teachers to engage in narratives of FoNs through different activities. The teachers read and listened to the stories proposed by the researcher, watched a narrative animated video, and staged FoNs Theaters (Corni and Fuchs 2021) related to the narratives. The different types of narratives were then discussed together. The second phase involved designing with the resources of the approach and testing them in the classroom. Anne experimented in fourth grade a narrative of FoN she elaborated with a group of four teachers and the support of the researcher. The narrative was told and mimed in class by the teacher. The students then took turns narrating the story and miming the characters, particularly the wind. The narration enabled aspects relating to a single FoN, the Wind, to emerge. In Italy, within the national indications for the curriculum, energy is only addressed in the fifth grade programme and therefore the decision was made to begin with a single FoN. Before fifth grade, FoNs aspects lay the foundations for understanding the concept of energy. Barbara and Cameron took a similar path, experimenting a narrative of FoNs in the fifth grade. "The Black Factory" story was developed by the researcher, inspired by the narratives of FoNs (Fuchs and Corni 2023), and was discussed together with the teachers before classroom practice. The story revolves around three young friends who fight to stop the production of tetraethyl lead in a factory and replace it with that of photovoltaic panels. In the narrative description of the panels' operation, interactions between FoNs emerge together with the concept of energy. Both teachers also implemented in class a connected FoNs Theater.

8 Data Sources and Methods of Data Collection

Data were collected during observations of the teachers' classroom practice and during subsequent individual interviews (Merriam 1998). As the focus was on the concept of energy, particular consideration was given to Barbara and Cameron's classroom practice of storytelling. The focus was on the linguistic expressions used

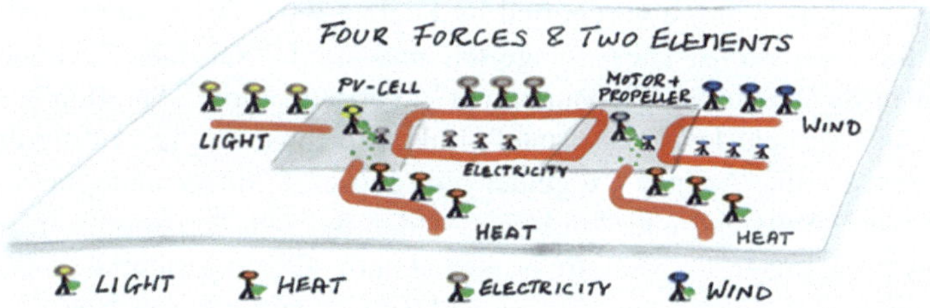

Fig. 1 Forces of Nature Theater (see Corni and Fuchs 2021 for a comprehensive description). Actors embody FoNs interacting and exchanging energy, represented by the green dust, on the stage, i.e., the technical systems. The red lines are paths followed by FoNs, while the size of the FoNs represents their high or low intensity

by the teacher in dialog with the students explaining and describing the narrative and related theater. In the individual interviews, all teachers explained and described narratively the FoNs interactions in the FoNs Theater, starting with a graphic representation related to the solar panel system, in the case of Barbara and Cameron (see Fig. 1), or in that of the wind turbine, in the case of Anne. For example, it was asked: "How would you explain in your own words to students what happens in this theater?; Who are the characters?; How do they move and what do they do? Why?; How do the characters interact?".

Institutional ethical research approval was granted and all ethical protocols were strictly adhered to.

9 Data Analysis

The teachers' set of linguistic expressions were analyzed by applying qualitative content analysis (Kuckartz and Rädiker 2023). Specifically, across cases commonalities were explored, identifying the common categories that emerge most frequently in each of the cases. The categories were derived from image schema and conceptual metaphor theory (Lakoff and Johnson 1980, 1999; Johnson 1987; Lakoff 1987). The linguistic expressions produced by the teachers were then "disassembled" to reveal first the conceptual metaphors and then the underlying image schemas, like in the examples below:

Light *arrives* at the panel *handing over* energy to electricity. [Metaphor: LIGHT IS A FLUIDLIKE SUBSTANCE – Image Schema: Fluidlike Substance; Metaphor: ENERGY IS A (FLUIDLIKE) SUBSTANCE – Image Schema: Substance].

Selected image schemas from the field of cognitive linguistics (Johnson 1987; Hampe 2005; Talmy 2000) and reported by Fuchs (2007, 2013) were used for the analysis of the expressions. These image schemas are: i. "Substance", connected to those of "quantity", "amount", "containment", "carrier", "flow", "motion", "source-path-goal"; ii. "Vertical level" or "state", linked to those of "level difference" or "state difference"; iii. "Agency", together with those of "direct manipulation" and "causation"; iv. "Activate" and "make"; v. "Identity"; vi. "creation". In order to analyze how these image schemas are projected onto other experiential domains, the theory of conceptual metaphor was employed (Lakoff and Johnson 1980).

10 Results and Discussion

10.1 Common Categories Related to the Concept of Energy in Teachers' Metaphorical Expressions

From the analysis, several common energy-related conceptualizations, or categories, emerged. These are presented in Table 1 and discussed in the following paragraphs. Table 1 shows the energy conceptions common to all the three cases and the underlying conceptual metaphors and image schemas. Note that each conception underlies the first metaphor ENERGY IS A (FLUIDLIKE) SUBSTANCE. Progressing down the table, only the metaphors and the image schemas that were not addressed before are highlighted. Moreover, concepts not shared by all teachers are in brackets.

Energy is only dealt with in relation with Forces of Nature. Teachers talked about energy by relating it to FoNs, as the citations in the lines 2, 3, 4, 5, 7, 9 of the table show. Indeed, the aspects of FoNs and the interactions between them underpin the imaginative understanding of the concept of energy (Corni et al. 2019; Corni and Fuchs 2020; Fuchs et al. 2011). In order to talk about energy, teachers therefore recur to conceptual metaphors and underlying image schemas that characterize FoNs (Fuchs 2007, 2010).

Energy is a (fluidlike) substance. Expressions like "energy is like a kind of *fluid*" (A1), "energy is that *stuff*" (B1) or the question that Cameron poses to students "How would you *materially* represent energy?" (C1) underlie the metaphor ENERGY IS A (FLUIDIKE) SUBSTANCE (Amin 2009; Scherr et al. 2012; Lancor 2014; Harrer 2017; Fuchs and Corni 2023). This conceptual metaphor is obtained with the projection of the "(fluidlike) substance" schema, onto the experiential domain of energy. The concept of energy as a substance is also implicit in "The children representing electricity […] *hand over* the energy" (Cameron).

Energy is a (fluidlike) substance contained in FoNs and/or technical systems. FoNs are said to be "*full of* energy" (A2), "*discharged*" (C2) if they do not have

Table 1 Teachers' energy conceptions and underlying conceptual metaphors and image schemas

Energy conceptions, metaphors, and underlying image schemas	Metaphorical expressions Anne	Metaphorical expressions Barbara	Metaphorical expressions Cameron
1. Metaphor (M): ENERGY IS A (FLUIDLIKE) SUBSTANCE Image Schema (IS): (fluid) substance	"energy is like *a kind of fluid*" (A1)	"energy is that *stuff* that makes everything move" (B1)	"How would you *materially* represent energy?" (C1)
2. …contained in FoN M: FORCE OF NATURE IS A CONTAINER IS: container, collection, space (in–out)	"the acting FoN, wind, is *full of* energy" (A2)	"energy *is present* in FoNs" (B2)	"the 'electricity children' [the children representing electricity] returned *discharged* to take more energy from the light" (C2)
3. …flowing along a path carried by FoN M: FON IS AN ENERGY CARRIER IS: carrier, source-path-goal, flow, motion	"energy is this fluid that passes from one FoN to another" (A3)	"'Light children' carry energy which is handed over to electricity and therefore electricity moves […] electricity hands over its energy to motion…" (B3)	"[…] when the light arrives, the electricity activates and moves carrying the energy, walks through the wires and arrives at the fan" (C3)
4. …carried by FoN when they are tensed, at high potential M: INTENSITY (OF THE FON) IS A VERTICAL LEVEL/ STATE IS: vertical level/state	"the wind, after giving up its energy, *goes to zero* and leaves without the green bricks. It practically loses intensity" (A4)	"If we switch off the light the electricity is there but it *is asleep* and that energy is still" (B4.1)	"the children representing electricity *were seated* inside the solar panel and then *stood up*. Their *intensity raised* as the light and energy arrived"; "electricity goes back *tired*" (C4)
5. …causing effects M: ENERGY IS A POWERFUL AGENT IS: make, activate	"is a transfer of energy that *allows* to switch the various interconnected FoN on and off" (A5)	"energy is that stuff that *makes everything move*: motion, heat, light" (B5)	"energy is something that *activates* and *makes move*" (C5)
6. …made available by the acting FoN M: FON IS A POWERFUL AGENT IS: direct manipulation, causation, agency, power	"*the acting FoN, the wind*, is full of energy and then this energy […] *is* practically passed to the electricity FoN, initially at zero" (A6)	"the moment *light* arrives with the energy of the sun, the children representing electricity are activated and start moving fast" (B6)	"*light* arrives at the panel handing over energy to electricity" (C6)

(continued)

Table 1 (continued)

Energy conceptions, metaphors, and underlying image schemas	Metaphorical expressions Anne	Metaphorical expressions Barbara	Metaphorical expressions Cameron
7. …made available by the acting FoN (through a potential fall) M: INTENSITY DIFFERENCE IS A LEVEL/STATE DIFFERENCE IS: level/state difference	"energy is this fluid that passes *from one FoN to another* because there is a potential difference. On one side it [the FoN potential] *goes down* and on the other side it *goes up*" (A7)		"electricity *activates* and moves carrying the energy, walks through the wires and arrives at the fan. The fan receives the energy and activates. Electricity then goes back *tired* to take more energy" (C7)
8. …received and used by the patient FoN and causes a rise in the patient's potential M: FoN IS A PATIENT (SUFFERING EFFECTS) IS: patient	"*another FoN switches on* because its potential is raised by this energy exchange" (A8)	"do you remember that '*electricity children*' were tired and then, as soon as they get the energy, they *start moving*?" (B8)	"the '*electricity children*' were seated inside the solar panel *and then stood up*. Their *intensity raised* as the light and energy arrived" (C8)
9. …partly missed by the patient FoN and used by Heat M: HEAT IS A FLUIDLIKE SUBSTANCE IS: creation	"*A part of* this energy that is handed over by the wind is actually *collected by the heat*, it is *dispersed* in the form of heat" (A9)	"*part of the energy* is *passed on to heat* because heat is released through movement and is *dispersed* into the air"(B9.1); "All this *produces* heat" (B9.2)	"energy is something that *dissipates*" (C9)
10. …which is always the same, does not change form M: ENERGY IS A FLUIDLIKE SUBSTANCE IS: identity	"it is incorrect to say that wind energy is transformed into electricity" (A10)	"energy *does not transform*, it is a constant" (B10)	"energy *remains the same*, FoN changes"(C10)

energy and therefore "energy *is present in* Forces of Nature" (B2) or "*in* the solar panel" (Barbara). These metaphorical expressions refer to the concept of energy storage and are structured through the projection of the "container", "space (in–out)", and "collection" schemas onto the FoN domain (Corni and Fuchs 2020). The main conceptual metaphor is thus FoN IS A CONTAINER.

Energy is a (fluidlike) substance flowing along a path carried by FoNs. Energy is also transported by FoNs along a path. This concept is exemplified by Barbara "'Light children' carry energy which is handed over to electricity and therefore electricity moves […] electricity hands over its energy to motion…"(B3). The other teachers talk about FoN that "carries" energy (B3, C3), or say that energy "passes from one Force to another" (A3) and "handed over" (B3) to another FoN. These expressions imply the conceptual metaphors: i. FON IS A (ENERGY) CARRIER, elaborated from the image scheme "carrier", ii. FON (TOGETHER WITH ENERGY) IS A FLUIDLIKE SUBSTANCE (MOVING ALONG A PATH), originated from the schemas "flow", "motion", and "source-path-goal" (Fuchs 2007; Corni and Fuchs 2020).

Energy is a (fluidlike) substance carried by FoN when they are tense, at high intensity. Teachers also highlight that the FoN when transporting energy is tense. They report that electricity "moves *intensively*" when it receives energy (Barbara) or that the children representing electricity "were *faster*" after taking energy from light (Cameron). Specifically, the expressions "Wind *goes to zero*" (A4) and "the 'electricity children' *were seated* inside the solar panel and then *stood up*. Their *intensity raised* as the light and energy arrived" (C4) underpin the metaphor INTENSITY IS A VERTICAL LEVEL, obtained by projecting the image schema of "level" or "scale" on the intensive aspect of the Force (Fuchs 2007). In addition, this metaphor is also conveyed by the level of the children's position in space. The intensity of a force is in fact high with respect to another location in the physical system. Moreover, tensions drive processes (Fuchs and Corni 2023) as Barbara explains "if there is no force that takes this energy, the energy is still, is nothing. If we switch off the light, the electricity is there but it is asleep and its energy is still".

Energy is a powerful agent which causes effects. Teachers sometimes refer to energy by attributing to it the image schema of agency and creating the metaphor ENERGY IS A POWERFUL AGENT: "energy allows the transition from the *effects* of one FoN to the *effects* of another FoN" (Anne) or "energy *makes* everything move" (B5). In accordance with what is argued in the previous paragraph, i.e., that FoN-related tensions are primary and drive processes, it is incorrect to speak of energy as a powerful agent. Not only laypeople but also scientists in their research talk about energy as a cause (Corni et al. 2019) and teachers might have been influenced by it.

Energy is a (fluidlike) substance made available by the acting FoN (through a potential fall). The concept of energy made available by the initially tense, acting FoN through a fall in potential (Fuchs and Corni 2023) is exemplified in this expression "the *acting FoN*, the wind, is full of energy and then this energy […] is practically passed to the electricity FoN, initially at zero. […] The wind, *after handing over its energy*, obviously *goes to zero* and leaves without its small green bricks. It practically *loses its intensity*" (A6). The term "*acting* FoN" suggests the metaphor FON IS A POWERFUL AGENT, obtained from the image schemas of "direct manipulation", "causation", "agency". We further deduce that the wind at the beginning of the interaction has a high intensity, that then "*goes to zero*", as the interaction progresses. The speed difference is thus conceptualized through the metaphor INTENSITY DIFFERENCE IS A LEVEL DIFFERENCE, which underlies the image schema "level difference" (Fuchs 2007; Fuchs and Corni 2023). The concept of potential drop is intended by Cameron through the image schema "state difference", as electricity "goes back *tired*" (C7) after the interaction.

Energy is a (fluidlike) substance received and used by the patient FoN and causes a rise in the patient's potential. In their expressions, teachers always identify a FoN who is affected by the acting FoN: the FoN that "receives" the energy "handed over" by the first FoN and experiences the consequences (Corni and Fuchs 2020). Teachers hence talk about a FoN that "activates" (C3) or is "activated" (B6), "start moving" (B8), "switches on" (A8) or "moves intensively" (Barbara), as soon as it "receives" (B4.2) energy or as soon as energy "arrives" (B6). These narrative expressions hide the metaphor FoN IS A PATIENT (SUFFERING EFFECTS).

Energy is a (fluidlike) substance with a (conserved) quantity. The property of energy being a conserved quantity emerges from expressions such as "energy is conserved" and "energy is a constant quantity", by Anne and Barbara, respectively. Energy is indeed conserved, and we always find the same amount of energy in nature. However, the teachers refer to it, without elaborating on the concept, so it is difficult to understand what they really meant.

Energy is a (fluidlike) substance partly missed by the patient FoN and used to produce Heat. Teachers say that energy is "collected by heat" (A9) or "passed on to heat" (B9.1), or, more in general, that is "dispersed" (A9, B9.1) and "something that dissipates" (C9). To understand imaginatively this concept, two of the teachers say that energy is partly missed by a FoN and collected, accepted by Heat: "a part of this energy that is handed over by the wind (to electricity) is actually collected by the Heat" (A9). Lastly Barbara claims that the process "produces Heat" (B9.2), underpinning the "creation" image schema. Teachers therefore characterize Heat by all the typical aspects of a FoN (Fuchs et al. 2022). Anne's expression "energy is dispersed in the form of heat", however, could recall the concept introduced by Clausius (1850).

Energy is a (fluidlike) substance and does not change form. Teachers do not think of energy as having different forms but clearly differentiate it from the FoN that carries it (Corni et al. 2019; Fuchs and Corni 2023), saying that "energy *does not transform*" (B10) or "energy *is not the same as electricity*" (Barbara), "energy *remains the same*" (C10). Anne, for example, recognizes that "energy cannot be defined by adjectives" like "chemical, kinetic, electrical…" and therefore "is incorrect" to say that "wind energy is transformed into electricity". These expressions are structured by the metaphorical projection of the image schema "identity" onto the experiential domain of energy.

There are no substantial differences between the energy-related conceptualizations of the three teachers. It is noticeable that Anne and Barbara explain the conceptualization of energy in line 9 and line 10 of Table 1 in a more articulate form. Barbara, on the other hand, unlike the other two teachers, does not refer to the conceptualization in line 7. Interestingly, however, Barbara, who was already familiar with the approach at the beginning of the study, refers to the FoNs Theater in her classroom dialog with the pupils to support her explanations of the concepts (B8). In addition, in an intervention not reported in the table, Barbara, in explaining the concept in line 3 to the pupils, mimes a FoN by pretending to carry the pupils, who would represent energy.

10.2 Other Relationships Between Categories in Teachers' Metaphorical Expressions

The teachers appear to understand energy through an interrelation between different conceptual metaphors. In addition, the expressions of all the teachers revealed other relevant significant relationships: i. between potential difference and substance flow; ii. between potential difference, substance flow, and power. The latter is strongly linked to the concept of energy (Fuchs 2007; Corni and Fuchs 2020). The relationship between potential difference and substance flow takes shape in the following expressions: i. "a potential difference is established, electricity *becomes tense* and therefore *electric current is produced* "(Anne); ii. "The moment the light arrives with energy from the sun, the children representing electricity *are activated* and *start moving* fast. […] Electricity hands over its energy to motion […] (B6); iii. "[The children who represented light] *were sitting and then stood up to give the energy* to those inside the panel [who represented electricity]" (Cameron). In these expressions, it is clear how the initial tension of a FoN triggers the flow of substance and drives the entire process. The last two expressions include a third element: power. This relationship also emerges from the following expressions: i. "We have a *potential difference*, in the sense that initially the agent FoN, the wind is charged with energy and then this energy […] *is* practically *handed over* to the electricity FoN, which is initially at zero (Anne); ii. The *more light* there is, the *stronger* it is, the *more energy* is transported (Barbara). Teachers thus qualitatively express the concept that the power of a phenomenon is proportional to the flow of the quantity and the potential difference through which the quantity flows (Fuchs 2007; Corni and Fuchs 2020).

11 Conclusion

Teachers conceive energy as a (fluidlike) substance contained in FoNs and technical systems, flowing along a path carried by FoNs when they are tensed, at high potential. Moreover, their expressions indicated that they asserted that energy is exchanged during FoNs interactions: is made available by the acting FoN (through a potential fall) and received and used by the patient FoN, causing a rise in the patient's potential. According to the teachers' statements, furthermore this fluidlike quantity is not only exchanged between FoNs, but has also the power to cause effects, to make things move. Energy is also said to be a (conserved) quantity and to be partly missed by the patient FoN and used by Heat. Lastly, teachers say that energy does not change form. With the exception of the idea of energy as a powerful agent, the other conceptualizations are in line with the notion of energy proper to the Imaginative Approach to PPE (Fuchs and Corni 2023). We can therefore argue that, through the engagement with narratives of Forces of Nature, teachers have developed an articulated conception of energy, which includes connections between elements and sheds light on the different properties of the concept. Teachers' language shows rich occurrences of figurative structures interconnected between each other. We know, even before recent studies in cognitive linguistics, that metaphor is a cognitive tool with a heuristic power (Black 1962) and creative power (Ricoeur 2004), perspectives that enhance the role of metaphor in its linguistic and communicative dimension (Contini 2020). Teachers therefore elaborated their learnings in a personal, creative, and original way. This interweaving of metaphors gave rise to a more complex structure, which is that of their narrative explanations or descriptions. Narrative therefore proved to be a meaning-making tool (Bruner 1991; Egan 1997) that allowed participants to reflect on the meaning of the concept of energy.

12 Fundings

This research project is financed by the National Operational Programme (PON) Research and Innovation 2014–2020 (CCI2014IT16M2OP005), FSE REACT-EU resources, Action IV.5 "Doctorate on Green Themes".

Acknowledgements We thank the Erasmus+ project "e^4—tools in higher Education for an Embodied & creative Energy Education", cooperation partnerships in Higher Education, number: 2021-1-IT02-KA220-HED-000032259, because it was a source for inspiration and knowledge for this study.

References

Amin TG (2009) Conceptual metaphor meets conceptual change. Hum Dev 52(3):165–197. https://doi.org/10.1159/000213891

Black M (1962) Models and metaphors. Cornell University Press, Ithaca, NY

Brewe E (2011) Energy as a substancelike quantity that flows: theoretical considerations and pedagogical consequences. Phys Rev Spec Top-Phys Educ Res 7(2):020106. https://doi.org/10.1103/PhysRevSTPER.7.020106

Bruner JS (1990) Acts of meaning. Harvard University Press, Cambridge

Bruner J (1991) The narrative construction of reality. Crit Inq 18(1):1–21

Carnot S (1824) Réflexions sur la puissance motrice du feu et sur les machines propres a developer cette puissance. Bachelier, Paris

Clausius R (1850) Über die bewegende Kraft der Wärme und die Gesetze, welche sich daraus für die Wärmelehre selbst ableiten lassen. Annalen der Physik und Chemie 79(368–397):500–524

Corni F (ed) (2013) Le scienze nella prima educazione. Un approccio narrativo a un curricolo interdisciplinare. Erickson, Trento

Corni F, Fuchs HU, Landini A, Giliberti E (2019) Visual and gestural metaphors for introducing energy to student teachers of primary school and kindergarten levels. J Phys: Conf Ser 1287:012043. https://doi.org/10.1088/1742-6596/1287/1/012043

Contini A (2020) Black e Ricœur filosofi della metafora. Aisthema Int J 7(1):117–151. https://hdl.handle.net/11380/1225858

Corni F, Fuchs HU (2020) Primary physical science for student teachers at kindergarten and primary school levels: part I—Foundations of an imaginative approach to physical science. Interchange 51(3):315–343. https://doi.org/10.1007/s10780-019-09382-0
Corni F, Fuchs HU (2021) Primary physical science for student teachers at kindergarten and primary school levels: part II—Implementation and evaluation of a course. Interchange 52(2):203–236. https://doi.org/10.1007/s10780-021-09424-6
Delegkos N, Koliopoulos D (2020) Constructing the "energy" concept and its social use by students of primary education in Greece. Res Sci Educ 50(2):393–418. https://doi.org/10.1007/s11165-018-9694-y
Duit R (1987) Should energy be introduced as something quasi-material? Int J Sci Educ 9:139–145. https://doi.org/10.1080/0950069870090202
Egan K (1997) The educated mind. How cognitive tools shape our understanding. The University of Chicago Press, Chicago
Feynman R (1963) The Feynman lectures on physics. Book 1. Addison-Wesley, New York
Fuchs HU (1987) Entropy in the teaching of introductory thermodynamics. Am J Phys 55:215–219. https://doi.org/10.1119/1.15216
Fuchs HU (1996) The dynamics of heat. Springer, New York
Fuchs HU (2006) From image schemas to dynamical models in fluids, electricity, heat, and motion. Examples, practical experience, and philosophy. Invited talk. In: Proceedings of the 2006 GIREP conference. University of Amsterdam
Fuchs HU (2007) From image schemas to dynamical models in fluids, electricity, heat, and motion. Essay Phys Educ Res
Fuchs HU (2010) Force dynamic gestalt, metafora e pensiero scientifico. In: Corni F, Mariani C, Laurenti E (eds) Innovazione nella didattica delle scienze nella scuola primaria: Al crocevia fra discipline scientifiche e umanistiche. Artestampa, Modena, pp 8–18
Fuchs HU, Corni F, Giliberti E, Mariani C (2011) Force dynamic gestalt of natural phenomena: teaching the concept of energy. In: Proceedings of the biannual conference of the European science education research association ESERA 2011 CONFERENCE. Lyon, France
Fuchs HU (2013) Il significato in natura. In: Corni F (ed) Le scienze nella prima educazione. Un approccio narrativo a un curricolo interdisciplinare. Trento IT: Erickson. English version: meaning in nature: from schematic to narrative structures of science
Fuchs HU (2014) The narrative structure of continuum thermodynamics. In: Constantinou CP, Papadouris N, Hadjigeorgiou A (eds) E-book proceedings of the ESERA 2013 conference: science education research for evidence-based teaching and coherence in learning. Part 1. European Science Education Research Association, Nicosia, pp 50–51
Fuchs HU (2015) From stories to scientific models and back: narrative framing in modern macroscopic physics. Int J Sci Educ 37(5–6):934–957
Fuchs HU, Contini A, Dumont E, Landini A, Corni F (2018) How metaphor and narrative interact in stories of forces of nature. In: Narrative and metaphor in education. Routledge, pp 91–104
Fuchs HU, D'Anna M, Corni F (2022) Entropy and the experience of heat. Entropy 24(5):646. https://doi.org/10.3390/e24050646
Fuchs HU, Corni F (2023) Primary physical science education: an imaginative approach to encounters with nature. Springer Nature
Hampe B (2005) From perception to meaning. Image schemas in cognitive linguistics. Mouton de Gruyter, Berlin
Harrer BW (2017) On the origin of energy: metaphors and manifestations as resources for conceptualizing and measuring the invisible, imponderable. Am J Phys 85(6):454–460. https://doi.org/10.1119/1.4979538
Herman D (2002) Story logic. University of Nebraska Press, Lincoln, NE
Herman D (2009) Basic elements of narrative. Wiley-Blackwell, Chichester
Heron P, Michelini M, Stefanel A (2009) Teaching and learning the concept of energy at 14 years old. In: Bisel A, Garip M (eds) Proceedings of selected papers to the Frontiers in science education research conference, EMU, Famagusta Cyprus, pp 231–240

Hertting S (2016) Energy blocks—A physical model for teaching energy concepts. Phys Teach 54(1):31–33. https://doi.org/10.1119/1.4937969

Johnson M (1987) The body in the mind. University of Chicago Press, Chicago

Kuckartz U, Rädiker S (2023) Qualitative content analysis: methods, practice and software. SAGE Publications

Lakoff G, Johnson M (1980) Metaphors we live by. University of Chicago Press, Chicago

Lakoff G (1987) Women, fire, and dangerous things. University of Chicago Press, Chicago

Lakoff G, Johnson M (1999) Philosophy in the flesh. Basic Books, New York

Lancor R (2014) Using metaphor theory to examine conceptions of energy in biology, chemistry, and physics. Sci Educ 23(6):1245–1267. https://doi.org/10.1007/s11191-012-9535-8

Lee HS, Liu OL (2010) Assessing learning progression of energy concepts across middle school grades: the knowledge integration perspective. Sci Educ 94(4):665–688. https://doi.org/10.1002/sce.20382

Liu X, McKeough A (2005) Developmental growth in students' concept of energy: analysis of selected items from the TIMSS database. J Res Sci Teach 42(5):493–517. https://doi.org/10.1002/tea.20060

Maurer W (1990). Ingenieurphysik auf neuen Wegen. Technische Rundschau 82(29/30):12–16

Merriam SB (1998) Qualitative research and case study applications in education. Jossey-Bass, San Francisco, CA

Millar RH (2005) Teaching about energy. Report. University of York, Department of Educational Studies, York

Millar R (2014) Towards a research-informed teaching sequence for energy. In: Teaching and learning of energy in K-12 education. Springer International Publishing, Cham, pp 187–206. https://doi.org/10.1007/978-3-319-05017-1_11

Neumann K, Viering T, Boone WJ, Fischer HE (2013) Towards a learning progression of energy. J Res Sci Teach 50(2):162–188. https://doi.org/10.1002/tea.21061

Papadouris N, Constantinou CP (2011) A philosophically informed teaching proposal on the topic of energy for students aged 11–14. Sci Educ 20(10):961–979. https://doi.org/10.1007/s11191-010-9305-4

Papadouris N, Constantinou CP (2016) Investigating middle school students' ability to develop energy as a framework for analyzing simple physical phenomena. J Res Sci Teach 53(1):119–145

Ricoeur P (2004) The rule of metaphor: the creation of meaning in language. Routledge

Scherr RE, Close HG, McKagan SB, Vokos S (2012) Representing energy. I. Representing a substance ontology for energy. Phys Rev Spec Top Phys Educ Res 8(2):020114

Stake RE (1995) The art of case study research. SAGE Publications, Thousand Oaks, CA

Talmy L (2000) Toward a cognitive semantics. Volume I: concept structuring systems. The MIT Press, Cambridge, Massachusetts

Toli G, Kallery M (2021) Enhancing student interest to promote learning in science: the case of the concept of energy. Educ Sci 11(5):220

Trumper R (1993) Children's energy concepts: a cross-age study. Int J Sci Educ 15(2):139–148

Van Driel JH, Verloop N, De Vos W (1998) Developing science teachers' pedagogical content knowledge. J Res Sci Teach: Offic J Natl Assoc Res Sci Teach 35(6):673–695

A Multi-phase Research on Developing Proposals for Teaching the Properties of Matter in Early Scientific Education: Focusing on Teacher Practices

Ayçin Ünal, Marisa Michelini, and Lorenzo Gianni Santi

Abstract This research project in the field of physics education consists of three phases: conceptual reconstruction of the topic, literature review on learning difficulties and teacher education; evaluation of primary and secondary school teachers' teaching practices on the properties of matter and their roles in physics; and focus group interviews with nine teachers to deepen the results of previous studies. Findings from the interviews show that some teachers do not have sufficient knowledge about the properties of matter and their roles in physics. Teachers often focus on the phase or state of properties of matter, but they emphasize the need to clarify the nature and relationship between the fundamental properties of matter. Although they state that understanding the properties of matter forms the basis of science education, they cannot provide sufficient examples of how and in what way it should be taught. As a result, teachers need to improve their content knowledge about the properties of matter and be supported with active learning methods that support formal thinking.

Keywords Properties of matter · Teaching practices · Focus group interview

1 Introduction

In school science curricula, there is a research area that is often quite neglected at all levels: properties and material science with all its important subfields. This is a broad research area that is significant in both fundamental and applied research and has

A. Ünal (✉) · M. Michelini · L. G. Santi (Deceased)
Research Unit in Physics Education, DMIF, University of Udine, Udine, Italy
e-mail: aycinunal@gmail.com

M. Michelini
e-mail: marisa.michelini@uniud.it

L. G. Santi (Deceased)
e-mail: lorenzo.santi@uniud.it

A. Ünal
Mugla Science and Art Center, Mugla, Turkey

I. Testa et al. (eds.), *Connecting Physics Education Research and Practice*,
Challenges in Physics Education, https://doi.org/10.1007/978-3-031-86609-8_16

a strong interdisciplinary specificity (Cohen and Louie 2016). While international comparisons show that physics has made countless contributions to this research area (Michelini et al. 1996), a consistent vertical teaching proposal for education has yet to be developed. At the university level, there are often contributions to specific research lines with advanced content, such as properties of fluids, electrical conduction properties in solids, and optical and thermal properties of certain materials. There are also introductory texts (McKenzie 2023), fundamentals (Cohen and Louie 2016), and methods and concepts specific to this research area (Continentino 2021) available for university students. Suggestions for teachers or the general public are not lacking: an example of this is Princeton University's offering (https://pccm.princeton.edu/outreach, https://pccm.princeton.edu/education/psscmp). However, these are often contributions to popularization or content presented in an interdisciplinary manner and lack the global vision necessary for students to acquire the conceptual framework or competency in managing various aspects.

When looking at secondary school textbooks, the topic of material properties, if present, is often scattered across sections of different physics theories such as mechanics, thermodynamics, electrodynamics, and optics. In the introduction to the topic of measurement, system properties like length, volume, and mass are discussed, measurement units systems are debated, but the nature of properties: the system, state, and material characteristics are not discussed. However, we know that these allow for the contextualization of physical laws and are the foundation of chemical laws, as well as creating criteria for material analysis and separation techniques. There are also no suggestions on the underlying theory of material properties. In some cases, models of the studied properties are shown, but no role is given regarding their basis, strength, and limitations: they are presented as imaginary solutions to unresolved problems. Material properties appear as parameters in the general laws of different physics sections, such as density in Stevin's law, specific heat in basic heating laws, thermal expansion coefficient in the simplified relative law for solids and liquids, and refractive index in Snell's law. Very rarely, significant relationships between electrical, optical, or thermal properties of materials are reflected in phenomenological observations or better interpretive discussions, which have been an important research topic in physics for many years and are today subjects of further study (Guizzetti et al. 1990; Orsini and Salvatori 2023).

Exceptions include research-based educational proposals at high school level on the electrical properties of materials (Sconza et al. 1996; Fera et al. 2014), the relationship between thermal properties of materials and the study of thermal phenomena (Bosio et al. 1996), or laboratory-based studies of specific systems like a light bulb, which help students recognize electrical, thermal, and light emission properties in terms of energy (Bosio et al. 1996; Santi and Michelini 2007). An interesting educational approach involves a series of experiments related to liquid crystals (Cepic 2015) or nanotechnologies (Lisotti et al. 2013). Didactic research provides sufficient studies on learning difficulties and spontaneous ideas about the electrical properties of materials (Fera et al. 2014; Michelini and Vercellati 2011; Fera and Michelini 2012), exploratory suggestions on magnetic properties (Fedele et al. 2005; Michelini and Vercellati 2012), and thermal properties (Benciolini et al. 2002) for preschool and primary school children. There are also literature review studies on materials and their properties (Ünal et al. 2023).

When seeking reference materials for scientific education at the primary and secondary school levels, the main focus is on phenomena as: phase changes, conservation of matter in physical and chemical transformations, buoyancy, solutions, suspensions, colloids, melting, and freezing. Characterization of the phase/state of matter (solid, liquid, gas) is often covered. Another perspective we could find with a specific frequency is the particle model of matter. In general, the specific properties of matter, i.e., the clear distinction between physical systems, materials, and states, are lacking. Only a few qualitative videos address the question "What are the Special Properties of Matter?".[1] Few consistent proposals are found in didactic projects, addressing the topic from the perspective of material properties useful in dissection processes. Examples include MIT's IPS and PS2 projects developed for students aged 14–16, which recognize the particle nature of matter (IPS Group 1969). Some textbooks, particularly those focused on technological teaching, address material properties in technical or commodity terms.

In our research approach to physics education (Michelini et al. 2023), we identified the need for a conceptual contribution that provides teachers with a global vision of the conceptual and interpretive aspects of material and material properties. This involves a slow and progressive interpretation and association deepening that demonstrates the validation of properties through coherent vertical paths. This nature of a consistent educational approach relies on the restructuring of educational contents and emerges when combined with students' and teachers' ideas. Our research results show that vertical development with an exploratory approach and evidence-based ideas for interpretation, modeling, and formal thought is crucial for learning. The foundation of concepts is reinforced through critical analysis associated with the methods of the discipline, where evidence-based expressions are grounded in experimental foundations. From this perspective, it is necessary to develop a proposal that distinguishes the properties of systems, material, and state in terms of guiding principles defined since the world's first discoveries.

1.1 Literature Review

Kara and Kingir's (2022) findings that teaching the properties of matter using a model-based constructivist approach enhances students' conceptual understanding and their ability to construct model-based arguments, facilitating conceptual change. Stamenkovski and Zajkov (2017) found that students struggle to understand the concept of mass, often confusing it with density and weight, suggesting the need for curriculum and textbook revisions. Wilcox et al. (2022) designed a 3-day lesson plan for teaching the properties of matter through inquiry-based activities, finding that student participation in experiments contributes to a deeper understanding of matter

[1] The very introductory video is available in https://youtu.be/frfrH8tyvhc?t=147.

properties. Adams and Feagin (2017) implemented a lesson flow based on the 5E method to help students learn the extensive properties of matter. The evaluation at the end of the lesson revealed that students learned the topics related to the properties of matter at the level of application skills.

Furthermore, some studies have focused on investigating the effectiveness of various lesson plans, teaching methods, and techniques for teaching matter properties. Allen and Rogers (2015) designed an activity using the Claim, Evidence, and Reasoning (CER) framework to help primary school students understand the properties of matter, suggesting that this approach could deepen students' scientific understanding. Adams and Feagin (2017) proposed a lesson plan aimed at reinforcing the extensive properties of matter. Kácovsky (2019) designed ten different experiments to teach the thermal properties of matter to high school and younger students, sharing their implementations. Zorlu and Sezek (2020) found that combining collaborative learning and modeling-based instruction in teaching matter and thermal events improves students' academic achievements and scientific process skills. Kara and Kingir (2022) offered evidence of the effectiveness of a model-based science writing approach in teaching the properties of matter. Akkus and Doymus (2022) compared the effectiveness of the Reading Writing Presentation (RWP) and Concept Puzzle Method (JG) on academic achievement in seventh-grade students regarding the properties of matter with a control group. After implementing the study, they applied a post-test to students and found that both the RWP and JG groups statistically outperformed the control group. Avcı (2022) determined that the puzzle technique enriched with educational games positively affects the communication skills of prospective elementary school teachers in science and technology education classes. Samarapungavan et al. (2023) found that the use of technology-mediated and model-based research lessons with kindergarten students led to significant improvements in their learning.

Furthermore, Hadenfeldt et al. (2014) proposed a model analyzing student learning progression in existing research on teaching and learning the properties of matter.

In addition to these studies, there are also researches focusing on investigating the role of teacher education and pedagogical content knowledge in teaching the properties of matter. Harrell and Subramaniam (2014) identified misconceptions among pre-service teachers regarding density. Hanuscin et al. (2018) identified difficulties among elementary science teachers in understanding the particle properties of matter. Cisterna et al. (2022) found that teachers have limited experience in the properties of matter and struggle to relate them to specific teaching practices. Mikeska et al. (2017) developed an assessment tool to measure elementary science teachers' content knowledge, including the properties of matter, in three different units. They found that these assessment items effectively captured differences in teachers' content knowledge. Avcı (2022) found that using the Jigsaw technique alongside educational games can improve the communication skills of pre-service elementary school teachers and increase interest in science lessons.

However, despite existing studies focusing on specific groups, considering the importance of the subject and conceptual development, it is worth noting the lack of research taking a holistic view of the topic and a characterization based on experimental methods. Given the limited research on teacher pedagogical content knowledge related to teaching this subject, understanding teachers' content knowledge and pedagogical skills regarding the teaching of this topic and obtaining their views on teaching it is also important. Therefore, in this study, we aimed to provide a comprehensive understanding of the current state of teaching matter properties in primary and middle schools, focusing on teacher perspectives, practices, and the factors that influence them.

2 Teacher Practices

Firstly, we aimed to comparatively examine primary and middle school teachers' perspectives and practices regarding the teaching of matter properties. The research collected data through surveys from 195 science teachers in Italy and Turkey. The survey included multiple-choice questions and Likert-type questions. Data related to the Likert-type questions were analyzed using independent t-tests and paired-sample t-tests. The findings indicate differences in the subtopics and concepts taught by teachers on the properties of matter. Among the most commonly taught topics in both groups were mass, density, and thermal conductivity of matter. Changes of state (phase) in matter are often cited as matter properties, even when the characteristic temperature for each material is not mentioned. There are some similarities between Italian and Turkish teachers in terms of teaching methods and techniques. Both groups rely on observation, description of events, and activities related to students' daily experiences. However, Italian teachers emphasize discussion, reading textbooks, and explanation as fundamental strategies, while Turkish teachers focus on explaining events with examples and verbally explaining concepts and laws. The findings show that teachers' perspectives and practices are influenced more by the country's educational context than by the school level. The results underscore the need for further research to better understand factors that may affect teachers' perspectives and practices in teaching matter properties.[2]

After this, focus group interviews were conducted with teachers to gain a deeper understanding of their practices and applications. The focus group interviews were conducted via Microsoft Teams and were completed in approximately 60–90 min. Nine teachers participated in the focus group interviews. When presenting the findings, teacher names were coded as $T1$, $T2$ ... $T9$. Demographic information about the participating teachers is provided in Table 1.

Six female and three male teachers participated in the study, with their professional experience ranging from 12 to 35 years. Among the participating teachers, 5 have undergraduate degrees, 3 have master's degrees, and 1 has a doctoral degree.

[2] The detailed analysis of the second inquiry was an accepted presentation in the GIREP-EPEC 2023 Conference, and the relative paper is submitted for selected paper proceedings.

Table 1 Demographic information

Teacher	Gender	Classes taught	Teaching experience (year)	Degree
T1	F	2	19	Bachelor's degree
T2	F	3	12	Master's degree
T3	M	4, 5	19	Bachelor's degree
T4	F	5–8	16	Bachelor's degree
T5	F	6, 7	15	Master's degree
T6	M	6–8	15	Doctorate
T7	F	7	26	Bachelor's degree
T8	M	2	35	Bachelor's degree
T9	F	5	17	Master's degree

The responses given by the teachers when asked about which properties of matter they taught, along with the researcher's notes on these responses, are provided in Table 2.

Upon reviewing the provided answers, it can be observed that most teachers focus on the topics of states of matter, phase changes, mass, volume, and density. This can be interpreted as a strong emphasis on these concepts in the curriculum. However,

Table 2 Teachers' responses on properties of matter taught

	Teacher
Definition of matter	T1, T2
Qualitative properties of matter (rough, smooth, transparent, opaque, floating, sinking, solid, hard, etc.)	T1, T2, T7, T8
Distinguishing properties	T1, T6, T7
States of matter	T1, T2, T4, T6, T8, T9
Phase changes	T1, T4, T5, T6, T9
Mass	T1, T2, T3, T4, T7, T9
Volume	T1, T2, T3, T4, T7, T9
Density	T2, T3, T6, T7, T9
Inertia	T3, T7
Optical properties (light transmission, etc.)	T3, T8, T9
Elasticity	T3
Particulate nature of matter	T5, T6, T9
Compounds, mixtures, physics and chemistry change, acid, and bases	T5
Melting and boiling point	T7
Sound-related properties	T9
Electrical properties (insulators, conductors)	T9
Thermal properties (thermal conductivity, insulation, expansion)	T9

despite being included in the teaching programs, very few teachers reported covering topics such as compounds, mixtures, physical and chemical changes, acids and bases, melting and boiling points, sound-related properties, electrical properties (insulators, conductors), and thermal properties (thermal conductivity, insulation, expansion) in their lessons.

The responses given by the participating teachers regarding the importance of learning the properties of matter are provided in Table 3.

Most teachers agree that understanding the properties of matter is crucial for both scientific knowledge and a broader understanding of the world. However, concerns expressed by Teacher T7 emphasize the need for innovative teaching approaches to make these concepts more engaging and relevant to students.

The responses regarding how participant teachers relate the properties of matter to the laws of physics are given in Table 4.

As seen in Table 4, there are varying levels of understanding among teachers regarding how the properties of matter are related to different physical laws. While some teachers have a strong grasp of these concepts, others are less certain. Unfortunately, a significant number of teachers (T1, T5, T7) were unable to link the properties of matter to physical laws. This indicates a knowledge gap in fundamental physics concepts. The most frequently mentioned relationships involved mass and various

Table 3 Teachers' responses on the importance of learning the properties of matter

Theme	Teacher	Responses
Properties of matter and understanding the environment	T8, T9, T3, T2	"I think it is very important because when children understand matter, they gain a different perspective on the objects around them. By understanding its internal structure, they can realize what elements matter consists of."—T9
Foundation for science education	T1, T3, T6, T4	"Learning the properties of matter is important for my students. They need to learn it because it is the foundation of science education."—T1 "Learning the properties of matter is important for my students because it is the ABCs of this subject."—T3 "The properties of matter can be seen as one of the prerequisites in science education."—T4
Use in daily life	T8, T5, T4	"For example, knowing why a glass breaks or does not break based on its properties can be useful in life."—T8 "If students know the properties of matter, they can understand the properties of the materials they encounter daily."—T5
Student awareness and motivation	T7	"I don't think learning these properties is important for my students. These concepts seem a bit abstract for children. Most of the students, as we discussed last year, have forgotten most of the common and distinguishing properties. They don't seem very familiar with them and don't seem to care much about these distinguishing common properties."—T7

Table 4 Teacher responses about relating properties of matter to physics laws

Responses	Teachers
I don't know	T1, T5, T7
Mass-gravity law	T2, T9
Mass-Newton's laws	T4, T6
Mass-inertia	T2, T3, T6, T9
Particle structure of matter-chemistry-chemical reactions	T2, T4, T8
Mass-law of conservation of mass	T6
Particle structure of matter-solubility	T3
Optical properties-laws of light refraction	T4
Density-buoyant force	T8, T9
Particle structure of matter-chemical bonds	T9
Thermal properties-laws of thermodynamics and space studies	T8

physical laws. This suggests a strong understanding of the connection between mass and concepts such as gravity, Newton's laws, and inertia. A few teachers linked the particle nature of matter to chemical concepts like reactions, solubility, and bonds. This indicates a more focused understanding of matter at the atomic and molecular levels. Fewer teachers linked optical and thermal properties to specific physical laws. The inability of some teachers to relate the properties of matter to physical laws suggests a need for further professional development in these areas.

When we ask teachers to explain how the teaching and learning process unfolds when they teach the properties of matter in their lessons, some of the responses provided by teachers are given below.

T1: Firstly, I need to know what the child knows about the matter. In other words, I need to know what the child's knowledge accumulation is. Before teaching a concept, if you know what knowledge accumulation is, you do it with that. Especially with some questions, I try to measure the children's knowledge. What do you think they know and don't know about the matter? And especially in such cases, I bring materials to help them grasp the subject better. So, what are they doing? Through discovery, they determine the defining properties of matter by themselves. So, actually, by giving it in the form of a game, I make it smooth and seamless. Let's play a game. Do you think this is rough or smooth? I explain to them that after defining the matter with the sense organs, there are things that they can grasp. After explaining, I actually make children familiar with the subject through games using the five senses. I bring materials like plastic, hard, smooth, rough, and such, and, of course, after measuring the children's knowledge accumulation and grouping them according to what they know and don't know, I use gamification to make them discover one of the defining properties of each material by themselves. In this way, children discover the defining properties of matter through group work. They find them in the form of a game. Will it float or sink in water? Let's make paper boats or look, it's floating, it's sinking, is it transparent, translucent, opaque? What am I doing? I always ensure they discover through exploration. So, here it's not just about me telling them, but also about them discovering themselves, advancing by giving them the foundation of science. Discover for yourself that's the basis of science, you will find it, you will discover it. As I mentioned, in the third grade, we focus more on the defining properties of matter, in the second grade, we introduce it a bit, but in the fourth grade, we mainly touch upon topics like volume and mass, it continues in a spiral manner.

T2: We focus more on the child's near-to-far, concrete-to-abstract. Starting from what they see and know around them. For example, because we teach through movement and action, it becomes incredibly lasting. When teaching the properties of matter to children, you say it's hard, for example. It's a concept that's not forgotten or a concept like hardness. Because we, as Turkish people, are intelligent by sight, seeing, hearing, and touching are very important from near-to-far. We illustrate situations that can be perceived by the five senses, from concrete-to-abstract. For example, we present a drama by revealing situations that can be perceived by the five senses. We follow methods and techniques like videos, sounds, videos, games, and dramas more to visualize it for the child when teaching. When explaining the states of matter, almost the same, experiments can often be involved here.

T7: We don't have a laboratory or materials, so I only demonstrate with animations. How is the mass of a substance calculated, and how is it measured, how is volume measured? Well, we have graduated cylinders, so we can show volume, but I can't show mass because we don't have a scale. Well, then, the density is calculated by dividing this by that. We calculate the density of solids this way and, for liquids, we again use graduated cylinders. We don't measure mass directly, but I use animations. I also give some examples from daily life regarding the density of gases. Like, I show helium balloons. Or why carbon dioxide is used in extinguishing fires. Things like that. We talk about how the density of water decreases as it freezes, and so on. If I have the means, I conduct experiments, if not, I use online platforms, videos, and animations, I always demonstrate these virtually with experiments. Regarding electrical properties, they observe the conductivity or not of materials, they observe that the resistance of semi-metals is different. By using different metals and bulbs in a circuit, they see which ones conduct electricity better, and which ones conduct worse, based on the brightness of the bulb.

As a result, we can say that teachers use various methods based on their familiarity with the content and the available resources. Each teacher adapts their teaching methods according to their own conditions and the needs of their students, showing that the teaching process can be enriched with flexible and diverse approaches. In particular, there is a lack of a global vision in the scientific education goals in this topic area.

3 Discussion and Conclusions

Table 1 shows that the group of teachers interviewed is sufficiently representative of teachers working at the school level under study. The teachers list the properties of surface, material, and condition together without making a precise definition. As can be seen in Table 2, the teachers emphasize topics such as "States of matter," "Phase changes," "Mass," "Volume," and "Density" in their lessons. This shows that the curriculum places a strong emphasis on these concepts (Ministry of National Education of the Republic of Turkey 2018a; Ministry of National Education of the Republic of Turkey 2018b; Italian Ministry of Education, University and Research 2012). However, although these topics are included in the curriculum (Ministry of National Education of the Republic of Turkey 2018a; Ministry of National Education of the Republic of Turkey 2018b; Italian Ministry of Education, University and Research 2012), very few teachers reported that they covered the topics "Compounds," "Mixtures," "Physics and chemistry changes," "Acid and bases," "Melting and boiling points," "Sound-related properties," "Electrical properties (insulators, conductors)," and "Thermal properties (thermal conductivity, insulation, expansion)" in their lessons. In addition, teachers often address the topics of volume and mass together with density, but do not mention that the first two topics are properties of systems and the last one is properties of substances or materials.

As shown in Table 3, the majority of teachers emphasize that learning the properties of matter is important for understanding the world and for establishing a solid foundation in science education. However, the concerns expressed by Teacher T7 suggest that innovative teaching approaches are needed to make these concepts more engaging for students. The literature highlights that a variety of teaching methods and techniques can be used successfully in this context (Kara and Kingir 2022; Adams and Feagin 2017; Allen and Rogers 2015; Kácovský 2019; Zorlu and Sezek 2020; Akkus and Doymus 2022; Avcı 2022; Samarapungavan et al. 2023).

As seen in Table 4, the optical, elastic, and solubility properties of materials are addressed in an effort to relate material properties to the laws of physics. The responses indicate that some properties, such as inertia, are considered as properties of matter, a microscopic view of the structure or model of matter, and in terms of chemical categories such as mixtures and compounds. However, teachers do not exemplify properties related to the laws of physics, such as elasticity, resistance, refractive index, and specific heat. In addition, some teachers (T1, T5, T7) have difficulty in relating the properties of matter to the laws of physics, indicating a gap in understanding basic physics concepts. Interviews reveal that teachers have difficulty in establishing connections between the properties of matter and that some of the concepts they describe are not actually properties of matter. It appears that teachers do not have a sufficient understanding of the properties of matter and their role in physics. These findings indicate that teachers need to improve their content knowledge on the subject, and previous research confirms these results (Harrell and Subramaniam 2014; Cisterna et al. 2022).

Teachers believe that understanding the properties of matter improves students' understanding and interaction with their environment and forms the basis of science education, and also facilitates practical application in daily life. However, they have not explained what these properties are and how they are exemplified. Teachers generally adopt learning methods based on discovery and inquiry, but they develop a concept-based pedagogical vision on the properties of matter. This vision aims to encourage discovery by taking into account children's perceptions and ideas. While many teachers emphasize the importance of the technical benefits of material properties, some teachers state that the subject is not suitable for primary school levels and that children forget the properties. The responses reveal the lack of a global vision and epistemic role of teachers. In addition, teachers see students' lack of interest in the properties of matter as a problem and emphasize that concrete examples and activities can be useful. This situation points to a methodological need: the necessity of scientific approaches based on discovery and conceptualization. It is known that there are many methods and techniques for teaching the properties of matter in the literature (Kara and Kingir 2022; Adams and Feagin 2017; Allen and Rogers 2015; Kácovský 2019; Zorlu and Sezek 2020; Akkus and Doymus 2022; Avcı 2022; Samarapungavan et al. 2023). This shows that teachers need pedagogical content knowledge related to teaching this topic. The responses reveal that teachers generally adopt a hands-on learning approach when teaching the properties of matter. This approach includes activities and materials that require active student participation and facilitates experiential learning, but the properties and facts are treated on the same level. Teachers indicate that they adopt a concrete-to-abstract approach, starting with concrete objects and moving on to learn abstract concepts. The responses show the diversity of learning environments, teaching methods, materials, and assessment tools that teachers use to teach the properties of matter. This diversity shows that teachers adapt to different classroom conditions and students' diverse learning needs. It is important to consider educators' perceptions and beliefs about science and science education (Eberle 2008). Because teachers change the curriculum in a planned way with their individual teaching methods and belief systems (Cronin-Jones 1991; Schmidt et al. 1996). Therefore, it becomes clear that teachers need support at an interdisciplinary and epistemological level and a clear understanding of the properties and different roles of systems, substances, and states. They also need to have a global vision that recognizes that materials are parameters in many physical laws and that states are key to equilibrium conditions and transformations. In conclusion, our study highlights the need for teachers to develop a deeper understanding of the properties of matter and their roles in physics. Teachers should focus on clarifying the distinctions between different types of properties and their relationships: for this scope an educational reconstruction of the subject matter appears useful.

References

Adams K, Feagin S (2017) Describing matter. Sci Child 54(8):52

Akkus A, Doymus K (2022) Effect of subject jigsaw and reading writing presentation techniques on academic achievement of 7th grade science students' academic success in structure and properties of matter. Acta Didactica Napocensia 15(1):92–104

Allen J, Rogers MP (2015) Putting ideas on paper. Sci Child 53(3):32

Avcı F (2022) Effects on primary school teacher candidates of developing and implementing jigsaw technique activities enriched with educational games in science and technology teaching lessons. Acta Educationis Generalis 12(1):41–73

Benciolini L, Michelini M, Odorico A (2002) Formalizing thermal phenomena at 3–6 year old: action research in a teacher training activity. In: Michelini M, Cobal M (eds) Developing formal thinking in physics, Girep Book of selected papers. Forum, Udine, pp 391–396

Bosio S, Michelini M, Santi L (1996) From an incandescent lamp to the electrical properties of tungsten. In: Teaching the science of condensed matter and new materials, GIREP-ICPE Book. Forum, Udine, p 216

Cepic M (2015) Liquid crystals through experiments. IOP concise physics. Morgan & Claypool Publication

Cisterna D, Bookbinder AK, Mikeska JN, Lakhani HR (2022) Elementary preservice teachers' perceptions of assessment tasks that measure content knowledge for teaching about matter. J Sci Teacher Educ 33(8):910–937

Cohen ML, Louie SG (2016) Fundamentals of condensed matter physics. Cambridge University Press, Cambridge, Cambridge

Continentino MA (2021) Key Methods and concepts in condensed matter physics. IOP Publishing, Bristol

Cronin-Jones LL (1991) Science teacher beliefs and their influences in curriculum implementation: two case studies. J Res Sci Teach 28:235–250

Eberle F (2008) Teaching and coherent science: an investigation of teachers' beliefs about and practice of teaching science coherently. Sch Sci Math 108(3):103–112

Fedele B, Michelini M, Stefanel A (2005) 5–10 years old pupils explore magnetic phenomena in cognitive laboratory (CLOE), In: Pitntò R, Couso D (eds) Proceedings of the fifth international ESERA conference on contributions of research to enhancing students' interest in learning science. Barcelona

Ministry of National Education of the Republic of Turkey (2018a) Science Curriculum. http://mufredat.meb.gov.tr/Dosyalar/201812312311937-FEN%20B%C4%B0L%C4%B0MLER%C4%B0%20%C3%96%C4%9ERET%C4%B0M%20PROGRAMI2018.pdf. Accessed 07 May 2023

Ministry of National Education of the Republic of Turkey (2018b) Secondary Education Physics Curriculum. http://mufredat.meb.gov.tr/Dosyalar/201812103112910-orta%C3%B6%C4%9Fretim_fizik_son.pdf. Accessed 07 May 2023

Fera G, Michelini M (2012) Pupils' ideas exploration on metal electrical transport models in the informal context of a hands-on exhibit. Latin-Am J Phys Educ 6(supp. 1):198–207

Fera G (2014) Learning knots on electrical conduction in metals. In: Burra GS, Michelini M, Santi L (eds) Frontiers of fundamental physics and physics education research. Springer, Cham, Heidelberg, NY, Dordrecht, London, pp 561–566

Guizzetti G, Mazzega E, Michelini M, Nava F, Borghesi S, Piaggi A (1990) Electrical and optical characterization of GdSi2 and ErSi2 alloy thin films. J Appl Phys 67(7):3393–3399

Hadenfeldt JC, Liu X, Neumann K (2014) Framing students' progression in understanding matter: a review of previous research. Stud Sci Educ 50(2):181–208

Hanuscin DL, Cisterna D, Lipsitz K (2018) Elementary teachers' pedagogical content knowledge for teaching structure and properties of matter. J Sci Teach Educ 29(8):665–692

Harrell PE, Subramaniam K (2014) Teachers need to be smarter than a 5th grader: what pre-service teachers know about density. Electron J Res Sci Math Educ 18(6)

Italian Ministry of Education, University and Research (2012). National curriculum guidelines for pre-school and first cycle education. https://www.miur.gov.it/documents/20182/51310/DM+254_2012.pdf/1f967360-0ca6-48fb-95e9-c15d49f18831?version=1.0&t=1480418494262. Accessed 07 May 2023

IPS Group (1969) College introductory physical science and physical science, 2 edn. Prentice-Hall

Kácovský P (2019) The visit of a thermal imaging world in one physics lesson. Phys Educ 54(4):045011

Kara S, Kingir S (2022) Implementation of the model-based science writing heuristic approach in elementary school science. Int J Sci Math Educ 20(4):683–703

Lisotti A, De Renzi V, Rozzi CA, Villa E, Albertini F, Goldoni G (2013) Educational pathways through nanoscience: nitinol as a paradigmatic smart material. Phys Educ 48:298–311

McKenzie RH (2023) Condensed matter physics: a very short introduction, very short introductions, online edn. Oxford Academic, Oxford. https://doi.org/10.1093/actrade/9780198845423.001.0001. Accessed 14 May 2024

Michelini M, Vercellati S (2011) Primary pupils explore the relationship between magnetic fields and electricity. GIREP-EPEC & PHEC, pp 162–170

Michelini M, Pugliese S, Cobai D (1996) Teaching the science of condensed matter and new materials. GIREP, ICPE, ICTP editors. Udine, Forum

Michelini M, Santi L, Stefanel A, Chiofalo M (2023) Entangled research methods for the building of coherent conceptual thematic learning paths and connecting research with praxis. Il Nuovo Cimento 46(C):196

Michelini M, Vercellati S (2012) Pupils explore magnetic and electromagnetic phenomena in CLOE labs. Latin-Am J Phys Educ 6(supp. 1):10–15

Mikeska JN, Phelps G, Croft AJ (2017) Practice-based measures of elementary science teachers' content knowledge for teaching: initial item development and validity evidence. ETS Res Rep Ser 2017(1):1–72

Orsini A, Salvatori S (2023) electrical, optical, and transport properties of semiconductors. Nanomaterials 13:2615

PCCM. https://pccm.princeton.edu/outreach, https://pccm.princeton.edu/education/psscmp. Accessed 14 May 2024

Samarapungavan A, Bryan LA, Staudt C, Sapkota B, Will Pinto HE, Broadhead JM, Kimball N (2023) Using technology-mediated inquiry to help young learners reimagine the visible world through simple particle models. J Res Sci Teach 60(2):390–422

Santi L, Michelini M (2007) An interpretative model to evaluate the thermal conductivity of the filament in a bulb lamp. In: Proceedings GIREP conference2006: modeling in physics and physics education, pp 811–814

Schmidt WH, Jorde D, Cogan LS, Barrier E, Gonzalo I, Moser U, Shimizu T et al (1996) Characterizing pedagogical flow: an investigation of mathematics and science teaching in six countries. Kluwer Academic Publishers, New York, USA

Sconza A, Torzo G, Delfitto G, Michelini M (1996) Electrical conduction in metals: a didactical experimental study. In: Teaching the science of condensed matter and new materials. Forum Editrice Universitaria, Udinese, pp 122–134

Stamenkovski S, Zajkov O (2017) Seventh grade students' qualitative understanding of the concept of mass influenced by real experiments and virtual experiments. Euro J Phys Educ 5(2):6–16

Ünal A, Santi LG, Michelini M, Taşar MF (2023) The literature review on teaching and learning the properties of matter in physics education. In: The 15th conference of the European science education research association (ESERA), Cappadocia, Turkey

Wilcox J, Reiter R, Rose A, Alberts A, Murano K (2022) Patterns that matter: helping second graders explore patterns of properties of matter and the nature of science. Sci Child 59(3):30–36

Zorlu Y, Sezek F (2020) An Investigation of the effect of students' academic achievement and science process skills application together with cooperative learning model and the modeling-based teaching method in teaching science courses. Int J Prog Educ 16(4):135–157

Designing a Teaching/Learning Sequence to Train In-Service Lower Secondary School Teachers in Physics

Matteo Tuveri, Arianna Steri, and Viviana Fanti

Abstract This paper presents the design of a teaching-learning sequence aimed at training lower secondary school teachers in physics. The study seeks to provide a tool to enhance physics instruction and develop teachers' pedagogical content knowledge (PCK). We outline our theoretical framework, primarily grounded in Inquiry-Based Learning pedagogy, and explore its potential application in class to enhance teachers' professional development. Our findings highlight the importance of addressing mental models and misconceptions to promote targeted actions that strengthen PCK and facilitate conceptual change in teacher training.

1 Introduction

Teacher professional development remains a significant challenge in science education (Michelini and Sperandeo Mineo 2014b; Kulgemeyer and Riese 2018; Couso et al. 2022; Guisasola and McLoughlin 2022). According to the TIMSS survey (Mullins and Martin 2008), nearly all science teachers depend heavily on textbooks, with more than half focusing predominantly on theoretical explanations or exercises. Only 30% conduct in-class or laboratory experiments, and even fewer engage students in hands-on exploration. Effective curricula demand the reworking of scientific knowledge for pedagogical purposes, informed by historical, scientific,

M. Tuveri (✉) · V. Fanti
Dipartimento di Fisica, Università di Cagliari and INFN Sezione di Cagliari - Cittadella Universitaria di Monserrato, Monserrato, Italy
e-mail: matteo.tuveri@ca.infn.it

V. Fanti
e-mail: viviana.fanti@ca.infn.it

A. Steri
Dipartimento di Ingegneria Meccanica, Chimica e dei Materiali, Università di Cagliari, Cagliari, Italy
e-mail: arianna.steri@ca.infn.it

INFN Sezione di Cagliari - Cittadella Universitaria di Monserrato, 09042 Monserrato, Italy

I. Testa et al. (eds.), *Connecting Physics Education Research and Practice*,
Challenges in Physics Education, https://doi.org/10.1007/978-3-031-86609-8_17

and research-based principles of didactics and learning theory (Tiberghien et al. 2005; Leite 2002; Kulgemeyer et al. 2000).

Research consistently shows that student success in science education hinges on the quality of teacher preparation (McDermont 1990; National Research Council 1996, 2000; Burrows and Slater 2015; Caton 2021; Martins and Baptista 2024). Gess-Newsome et al. define high-quality teaching as the "purposeful and deliberate planning of instruction that results in a coherent learning experience for students" (Gess-Newsome et al. 2019, p. 945). As noted by the authors, to reach this purpose "a teacher must justify the importance and structure of the content they are teaching, their selection of pedagogical techniques, and how this combination effectively attends to the teaching context as it relates to meeting student-learning needs (background knowledge, preconceptions, and common areas of conceptual difficulty)". During this practice, teachers would learn how to create a respectful learning environment, assessing students' skills, or reviewing homework. More importantly, teachers would learn to develop more general and adaptable skills of practice through their engagement in their profession. This process relies on a purposeful and deliberate planning of instruction that results in a coherent learning experience for students. This means that teachers should be trained to teach physics not as a finished body of precompiled concepts and principles but so that students could infer such principles through experimentation, argumentation, hypothetico-deductive reasoning, modeling, and other practices that are germane to the physics enterprise (Vokos et al. 2022).

In Italy, lower secondary school is the weakest sector of the education system (Mincu 2015; Masci et al. 2018). Many science teachers do not have a background in physics; instead, they hold degrees in chemistry, biology, natural sciences, geology, or mathematics. As a result, they often avoid addressing physics content and rely on traditional, lecture-based methods (Mincu 2019). There is an urgent need for targeted professional learning in physics, particularly in both content knowledge and pedagogical methodologies. Teachers must develop instructional content knowledge and cultivate motivation and positive attitudes toward teaching (Couso et al. 2022; Etkina and Planinsic 2024). Therefore it is essential to provide teachers with professional development opportunities that strengthen their skills, knowledge, and experience, thereby increasing their ability and preparedness to teach through integrated STEM and STEAM (Science, Technology, Engineering, Arts, Mathematics) approaches (Caton 2021; Martins and Baptista 2024; Li et al. 2020).

Since Shulman's groundbreaking work (Shulman 1986, 1987), Pedagogical Content Knowledge (PCK) has become a core component of teacher education programs (Halim and Subahan 2002; Michelini et al. 2014; Michelini and Stefanel 2014a). Although numerous definitions and conceptualizations of PCK exist (Guess-Newsome 1999; Park and Oliver 2008; Kind 2009), Shulman's original concept frames it as an integration of content and pedagogy that informs how teachers translate subject matter into effective instructional strategies. This includes teachers' capacity to help students understand complex subject matter (Karal and Nedim 2016). What distinguishes teachers from other professionals is their ability to transform content knowledge into forms that are pedagogically impactful, engaging, and suited to diverse student populations (Shulman 1987; Halim and Subahan 2002). PCK thus offers a theoretical lens through which to evaluate and improve teachers' professional skills, elevate their professional identity, and guide their educational development.

To effectively improve physics teachers' PCK at the lower secondary level, professional learning programs must prioritize methodologies that clarify how physics teachers' content knowledge, teaching experience, and personal beliefs contribute to their pedagogical effectiveness (McDermott and Shaffer 2000; McDermott 2006; Michelini and Stefanel 2015; Etkina 2015; Schiering 2022). At this educational level, physics is often confused with technology, taught as a deterministic, abstract discipline with limited practical relevance (Olsen et al. 2011). However, students are more likely to engage when they encounter real-world problems that inspire intellectual curiosity and critical thinking (Viennot 2008).

In this study, we attempted to create an instrument to enhance in-service lower secondary school teachers' professional learning, promoting physics literacy, and strengthening their PCK by offering a methodology that utilizes real-world, low-cost physics experiments. The tool was developed by the Physics Education Research group at the University of Cagliari, Italy, in collaboration with the Cagliari Division of the National Institute for Nuclear Physics (INFN). It is focused based on teaching-learning sequence pedagogy. The methodology, tested over a four-year trial period (2019–2022), was fully implemented in 2023 in the case of classical mechanics as part of the "AggiornaMenti" INFN course (Fanti et al. 2022). This paper outlines our theoretical framework and discusses our methodology focusing on its educational value in enhancing teachers' physics literacy and PCK. The outcomes of this project and its implementation during the aforementioned course will be detailed in a forthcoming paper.

2 Theoretical Framework and Educational Tools

For a long time, teacher education has predominantly focused on pre-service teachers' knowledge and dispositions. Recent studies (Couso et al. 2022; Burrows and Slater 2015; Caton 2021; Martins and Baptista 2024; Vokos et al. 2022) highlight the importance of teachers' preparation, particularly concerning their teaching practices (Feiman-Nemser and Remillard 1995). It is essential for researchers to further investigate the connection between professional knowledge and classroom actions, as professional knowledge underpins the academic training of future teachers (Liepertz and Borowski 2018).

Research has provided valuable insights into how professional knowledge influences teachers' classroom actions (Guisasola and McLoughlin 2022). Understanding this relationship remains vital, as professional knowledge and practice are foundational to teacher education. To address this, a shared model of professional knowledge was established, integrating the original concept of PCK with its contemporary application in classroom practices and its impact on student achievement. The initial attempt to formalize this was the Consensus Model (CM) of PCK created in

2012 (Gess-Newsome 2015). More recently, the Refined Consensus Model (RCM) was developed to resolve discrepancies in PCK components and terminology, aiming to foster a unified understanding of science teachers' PCK (Carlson et al. 2019; Schiering et al. 2023).

Building on the PCK components of earlier models, RCM posits that factors such as teachers' CK, teaching experience, and beliefs about teaching and learning science are crucial for the development of their PCK (Behling et al. 2022; Mientus et al. 2022). The RCM delineates various levels of PCK through concentric circles (Carlson et al. 2019). The outer circle represents foundational professional knowledge, including CK. The next circle, collective PCK (cPCK), encompasses publicly shared knowledge among educators. Personal PCK (pPCK) pertains to individual teachers' knowledge and skills applied in planning, teaching, and reflecting. Enacted PCK (ePCK) involves the application of pPCK in specific lessons for particular students. The exchange of knowledge among these levels supports the development of pPCK, influenced by teaching experience and teachers' beliefs (Schiering et al. 2023). The close linkage between the realms of PCK and their mutual influence is a continuum of discipline-, topic-, and concept-specific PCK that is described for cPCK and applies to all realms of PCK (Carlson et al. 2019). Considering the realms of PCK as a continuum implies that, horizontally, PCK ranges from a broader angle of view (discipline-specific) to a narrower angle of view (concept-specific). Vertically, the knowledge level of PCK, and especially pPCK and ePCK, can vary from the basic level to the expert level (Behling et al. 2022).

An effective approach to operationalizing PCK within the framework of RCM in a classroom setting is to design appropriate educational pathways based on the Model of Educational Reconstruction (MER). According to MER, the structure of a physics lesson differs from the structure of physics as a discipline itself (Duit et al. 2012). Lessons often rely on multiple representations of physics knowledge, such as those found in textbooks. However, these representations are typically abstract and condensed, targeting experts or at least soon-to-be-experts. The lesson content structure should be derived from the structure of physics content, requiring teachers to transform the latter to make it accessible to students. This transformation involves a process of elementarization, based on three key aspects (Duit et al. 2012): (a) identifying the fundamental principles and phenomena of a topic that align with instructional goals to support student understanding; (b) reducing the complexity of content while preserving scientific integrity; and (c) designing learning processes that guide students from initial ideas to fully developed scientific concepts. To ensure the content is meaningful, teachers must contextualize the elementarized ideas, which necessitates not only a deep understanding of the scientific content but also insight into students' perspectives, including their preconceptions, interests, and self-concepts in science, to effectively facilitate their learning.

For the educational reconstruction process to be effective, teachers must have a solid understanding of the subject matter (CK). This process is also shaped by their personal knowledge and skills in planning, teaching, and reflecting (pPCK), as well as their collaboration with colleagues and educators (cPCK). Additionally, it is influenced by their knowledge of students' understanding and instructional methods (ePCK). Collectively, this constitutes teachers' PCK within the RCM framework (Liepertz and Borowski 2018).

According to MER, the content structure is directly influenced by teachers' decision making which requires their CK as well as PCK in the sense of RCM. In order to plan and teach physics in the real in-class situation, fostering a high-quality teaching process, the conceptualization of physics, which involves establishing relationships between knowledge elements and different modes of reasoning should be taken into account. To characterize these relationships, we refer to Ohlsson's epistemic tasks such as describing, explaining, arguing, predicting, critiquing, and defining (Ohlsson 1996). Teachers are expected to identify patterns, explore relationships between objects, develop theoretical frameworks, test predictions, and make informed decisions (Tiberghien et al. 2005). The theory is presented in relation to teachers' everyday experiences, from which their conceptions emerge. Teachers are tasked with applying the theory to specific problems.

Inquiry-Based Learning (IBL) and the "PEC methodology" help implement these principles in the classroom (White and Gunstone 1992; Thornton and Sokoloff 1999; Theodorakakos and Psillos 2010; Tuveri et al. 2023). These approaches involve posing stimulus questions to make predictions (P), conducting experiments or explorations (E), and comparing outcomes (C), promoting knowledge construction through active engagement (Minner and Levy 2010). Teachers participate in IBL activities individually or in small groups, often using cooperative learning strategies (Tuveri et al. 2023; Duran and Dökme 2016).

Semiotics plays a crucial role in this process (Tiberghien et al. 2005). By analyzing teachers' language, ideas, and reasoning, researchers can identify gaps in their understanding and take targeted actions to support conceptual refinement. These are fundamental elements of conceptual change, helping in bridging the gap between teachers' initial knowledge and the knowledge required for effective teaching (Buty et al. 2004). Additionally, addressing misconceptions is vital for improving PCK. Strategies to overcome misconceptions involve making the learner's ideas intelligible, plausible, and fruitful, with analogies aiding in the acceptance of new knowledge during cognitive conflict (Duit 2009; Kuczmann 2021).

3 Methodology

To investigate the evolution of teachers' pre-instructional conceptions toward a more scientific understanding, flexible research methods are essential. These methods should accommodate individual learning processes and allow for tailored interventions, while also supporting teacher professional development and enhancing PCK as outlined in RCM. Our methodology employs a well-established approach based on MER: the teaching/learning sequences (Marzari et al. 2023; Méheut and Psillos 2004; Steffe and Thompson 2000; Komorek and Duit 2004; Engelhardt et al. 2004).

This approach involves multiple learning cycles where researchers test hypotheses regarding how teachers learn and reason about specific subjects (Magnusson et al. 1999). Each learning cycle comprises three phases: exploration, concept introduction, and concept application. In the exploration phase, teachers engage in hands-on activities to investigate concepts. The concept introduction phase involves explaining and refining observations. Finally, in the concept application phase, teachers apply the newly introduced concepts to novel situations. Each cycle includes discussions of various experiments and phenomena requiring explanation. From an operational point of view, we recast these cycles in four phases as described in Table 1: forming expectations, providing individual or group explanations, modeling, and discussion. The researcher assumes multiple roles throughout these cycles, extending beyond mere classroom management.

In each cycle, the researcher fosters deep pedagogical, conceptual, and scientific reasoning during theoretical steps, activities, or experiments. This involves PEC steps discussed in the previous section: brainstorming on teachers' existing knowledge of a conceptual issue, setting expectations for phenomena or experiments, observing outcomes, and evaluating and comparing expectations with observations. Additional steps include analyzing suggestions and motivations related to teachers' expectations, modeling concepts, and facilitating discussions. This process aids in conducting

Table 1 Topics for the obligatory questions to ask during an experimental phase of a given learning cycle

Tutor-class interaction	
Phase	Questions
Expectations before the experiment	– Indicate what you expect to observe – How will the experiment go – Indicate in which way you reached this assessment
Collegial explanation after the experiment	– Describe the observations – Indicate which of the object's or structure's characteristics are responsible for its behavior – Indicate which forces and other influences are present – Indicate what happens while repeating the experiment
Modeling	– Describe the theoretical, global, and local behavior – Find analogous relationships with other objects or experiments
Discussion	– Indicate your developing new concepts – Indicate what your study results mean for certain physics concepts, to what extent must they be altered, expanded, or redefined – Indicate what changes your view of physics – The pedagogy we are using and its possible implementation in in-class science lessons

training activities and interacting with the class. Each learning cycle begins with an inquiry-based question about everyday phenomena. Teachers generate hypotheses, which are then tested through experiments conducted using low-cost, pre-prepared experimental kits.

The researcher may use these kits independently to discuss specific phenomena or involve participants in the activity, facilitating reasoning and discussion with the entire class. During this phase, the researcher acts as a guide, posing a series of essential questions as illustrated in Table 1 (see also Nguyen and Duong (2021)).

These questions embody the epistemic tasks outlined by Ohlsson (1996) discussed in the previous Section. They are designed to stimulate teachers' reasoning, allowing their conceptions and mental representations to emerge. The outcomes of these interactions provide insights into their levels of scientific understanding and PCK. In an inquiry-based learning (IBL) framework, teachers engage in asking questions, collecting, discussing, and evaluating data to address queries posed by the researcher during discussions (2023). Comparing phenomena with differing features (e.g., the fall of two objects with equal mass and different shapes of vice versa in classical mechanics) promotes conceptual change and cognitive conflict. When experimental outcomes diverge from expectations, the researcher investigates teachers' mental models and theories, comparing them with observations, and guiding teachers toward a proper description of the phenomenon. This process corresponds to the modeling phase, focusing on identifying dependent and independent variables and their relationships to develop a suitable model based on a given theory.

Hands-on activities are conducted in small teacher groups to foster cooperative learning. Teachers explore phenomena using experimental kits prepared by researchers and construct their own experiments with low-cost materials brought from home. This activity involves a minds-on approach, where teachers design their experiments, discuss issues related to the subject and theory, assess whether their experiments work and why, compare observations with expectations, and analyze differences and similarities between real-world and idealized experiments.

Introducing teachers to IBL is effectively achieved by collaboratively conducting open-ended experiments. Teachers discuss the physical principles behind phenomena and construct experimental setups using everyday, low-cost materials. This approach enhances teachers' ability to analyze tasks or phenomena scientifically, extending their teaching beyond mathematical formulas to include natural language explanations. It supports the development of conceptual and descriptive understanding.

At the end of the training, teachers reflect on the activities and connect them to conceptual issues and methodologies discussed. The researcher discuss with them about their teaching practices (pPCK), gather feedback on the feasibility of the experimental activities and the methodology's implementation in in-class scenarios. Assessment concerning the impact of the sequence on teachers' lesson planning and students' understanding (ePCK) follows. Researchers also provide teachers with materials such as slides, video lectures, and experimental objects relevant to the topic of study (e.g., springs, weight scales, and dynamometers in classical mechanics).

4 Discussion and Conclusions

We developed an educational tool aimed at training in-service lower secondary school teachers in physics, providing them with a resource to enhance their physics knowledge and teaching practices. Our primary goal was to improve teachers' PCK by focusing on both content and pedagogical knowledge aspects according to RCM and MER.

To address our research goal, we designed a teaching/learning sequence that fosters a bottom-up learning process. By starting from common ground—shared and familiar ways of thinking about the world—we developed teaching activities that create opportunities within teachers' conceptual frameworks to deepen their physics understanding and conceptual grasp of the topics presented.

A key focus of our methodology is on modeling, which is crucial for enhancing teachers' PCK. As teachers work through the questions in Table 1, they engage in experimental testing, hypothesis evaluation, and scaffolded construction or deconstruction of their beliefs and conceptions (Nguyen and Duong 2021). To avoid fostering misconceptions and instead promote accurate understanding, teachers must make their conceptual constructions explicit, as noted in Lijnse (2004). By building on prior knowledge and employing a scientific inquiry-based approach, teachers can start design lessons that transform preliminary teaching/learning sequences according to our methodology into effective instructional didactic units (Duit and Treagust 1998). This process involves the elementarization process described in our theoretical framework (Duit et al. 2012), as well as a "scientificalization" of everyday experiences, rather than simply delivering scientific knowledge as a ready-made product (Freudenthal 1991). We also propose that social interactions and cooperative learning strategies can play a pivotal role in fostering knowledge construction and facilitating conceptual change. To further explore potential teacher misconceptions, we suggest focusing on semiotic registers, particularly the use of language and reasoning when analyzing and describing phenomena (Tiberghien et al. 2005). Rather than relying solely on mathematical descriptions and formulas, researchers should guide teachers in developing their logical and physical reasoning skills.

The effectiveness of lower secondary school teachers' professional development also relies on aligning educational strategies with their teaching orientations. Magnusson et al. (1999) describe orientations as a way of conceptualizing science teaching. Orientations shape how teachers conceptualize science teaching, guiding the design of training programs to address key objectives such as student assignments, curriculum use, and assessment (ePCK). The discussion phase at the end of the sequence exemplifies the development of cPCK and aids in fostering both pPCK and enacted ePCK. While teachers often cite time constraints in implementing experimental methods, viewing each phase of the sequence (learning cycles) independently can promote dynamic knowledge exchange among PCK components (Carlson et al. 2019). Knowledge-based reasoning, which consists of description, explanation, and prediction, plays a crucial role in transforming pPCK into ePCK (Behling et al. 2022; Santagata and Yeh 2016). This reasoning process influences teachers' motivation, beliefs, and enthusiasm, ultimately affecting their teaching quality and adaptability in various classroom situations. Implementing our methodology in in-class activities can enhance teachers' reasoning skills and professional development, especially concerning their ability to adjust teaching practices accordingly.

The proposed design of a teaching/learning sequence offers a methodology to discuss textbook topics from various perspectives using hands-on activities with low-cost, homemade experimental devices. We suggest that it enhances teachers' content knowledge (CK) and helps them design lessons aligned with the structure of physics content to make it accessible to students. This process fosters teachers' motivation in STEM orientating them toward science teaching. It supports professional development by enhancing PCK in line with MER and RCM.

In conclusion, our teaching/learning sequence aligns with our research framework, supporting professional development from a research-based perspective and a strong connection between empirical research on teaching and learning processes and instructional design. However, while our tool can qualitatively monitor teachers' PCK, a quantitative study with pre- and post-activity tests could provide more insights into the program's effectiveness (Gess-Newsome et al. 2019; Vokos et al. 2022; Liepertz and Borowski 2018; Vogelsang et al. 2022). Our future investigations will focus on testing the methodology over time with different topics and in-class activities to evaluate its impact on teaching and student learning of physics.

References

Behling F, Foertsch C, Neuhaus BJ (2022) The refined consensus model of pedagogical content knowledge (PCK): detecting filters between the realms of PCK. Educ Sci 12(9):592

Burrows A, Slater T (2015) A proposed integrated STEM framework for contemporary teacher preparation. Teach Educ Pract 28:318–330

Buty C, Tiberghien A, Le Maréchal J-F (2004) Learning hypotheses and an associated tool to design and to analyse teaching-learning sequences. Int J Sci Educ 26(5):579–604

Carlson J, Daehler KR, Alonzo AC, Barendsen E, Berry A, Borowski A, Carpendale J, Chan KKH, Cooper R, Friedrichsen PJ, Gess-Newsome J, Henze-Rietveld I, Hume A, Kirschner S, Liepertz S, Loughran J, Mavhunga E, Neumann K, Nilsson P et al (2019) The refined consensus model of pedagogical content knowledge in science education. In: Hume A, Cooper R, Borowski A (eds) Repositioning pedagogical content knowledge in Teachers' knowledge for teaching science. Springer Singapore, pp 77–92

Caton JC (2021) Don't run out of STEAM! Barriers to a transdisciplinary learning approach. J STEM Teach Educ 56:4

Couso D, Fazio C, Jeskova Z. Work Group 5 position paper: strategies for pre-service physics teacher education. J Phys: Conf Ser 2297:012024

Duit R, Treagust DF (1998) Learning in science—From behaviourism towards social constructivism and beyond. In: Fraser BJ, Tobin KJ (eds) International handbook of science education, pp 3–25

Duit R (2009) Bibliography STCSE: students' and teachers' conceptions and science education. University of Kiel, Kiel, Germany

Duit R, Gropengiesser H, Kattmann U, Komorek M, Parchmann I (2012) The model of educational reconstruction: a framework for improving teaching and learning science. In: Dillon J, Jorde D (eds) Science education research and practice in Europe: retrospective and prospective. SENSE, Rotterdam, Boston, Taipei, pp 13–27

Duran M, Dökme I (2016) The effect of the inquiry-based learning approach on student's critical thinking skills. Eurasia J Math Sci Technol Educ 12:2887

Engelhardt PV, Corpuz EG, Ozimek DJ, Sanjay Rebello N (2004) The teaching experiment—What it is and what it isn't. AIP Conf Proc 720:157–160

Etkina E (2015) Millikan award lecture: students of physics. Am J Phys 83(8):669–679

Etkina E, Planinsic G (2024) The investigative science learning environment: a guide for teacher preparation and professional development. IOP ebooks

Fanti V, Andreotti M, Beraudo A, Bertelli S, Cibinetto G, Embriaco A, Marcellini S, Postuma I, Protti N, Servoli L, Signorelli G, Tagliente G, Tuveri M, Zavatarelli S (2022) AggiornaMenti: an INFN project for the education of junior high school science teachers. PoS (ICHEP 2022), 1009

Feiman-Nemser S, Remillard J (1995) Perspectives on learning to teach. In: Murray F (ed) The teacher educator's handbook. Jossey-Bass, San Francisco, pp 63–91

Freudenthal H (1991) Revisiting mathematical education. Kluwer, Dordrecht

Gess-Newsome J (2015) A model of teacher professional knowledge and skill including PCK: Results of the thinking from the PCK summit. In: Berry A, Friedrichsen PJ, Loughran J (eds) Teaching and learning in science series. Re-examining pedagogical content knowledge in science education, 1st edn. Routledge, New York, NY, pp 28–42

Gess-Newsome J, Taylor JA, Carlson J, Gardner AL, Wilson CD, Stuhlsatz MAM (2019) Teacher pedagogical content knowledge, practice, and student achievement. Int J Sci Educ 41(7):944–963

Guess-Newsome J (1999) PCK: an introduction and orientation. In: Guess-Newsome J, Lederman NG (eds) Examining PCK (S. 3–17). Dordrecht, Kluwer

Guisasola J, McLoughlin E (2022) In: Guisasola J, McLoughlin E (eds) Connecting research in physics education with teacher education, vol 3. I.C.P.E. Book

Halim L, Subahan MMM (2002) Science trainee teachers' pedagogical content knowledge and its influence on physics teaching. Res Sci Educ Technol 20(2):215–225

Karal LS, Nedim A (2016) Development of pre-service physics teachers' pedagogical content knowledge (PCK) throughout their initial training. Teach Dev 20(2):162–180

Kind V (2009) Pedagogical content knowledge in science education: potential and perspectives for progress. Stud Sci Educ 45(2):169–204

Komorek M, Duit R (2004) The teaching experiment as a powerful method to develop and evaluate teaching and learning sequences in the domain of non-linear systems. Int J Sci Educ 26(5):619–633

Kuczmann I (2021) Misconceptions in mechanics and their elimination. J Phys: Conf Ser 1929:012005

Kulgemeyer C, Riese J (2018) From professional knowledge to professional performance: the impact of CK and PCK on teaching quality in explaining situations. J Res Sci Teach 55(10):1393–1418

Kulgemeyer C, Borowski A, Buschhueter D, Enkrott P, Kempin M, Reinhold P, Riese J, Schecker H, Schroeder J, Vogelsang C (2020) Professional knowledge affects action-related skills: the development of preservice physics teachers' explaining skills during a field experience. J Res Sci Ed 57(10):1554–1582

Leite L (2002) History of science in science education: development and validation of a checklist for analysing the historical content of science textbooks. Sci Educ 11:333

Liepertz S., Borowski A (2018) Testing the Consensus Model: relationships among physics teachers' professional knowledge, interconnectedness of content structure and student achievement. Int J Sci Ed

Lijnse P, Klaassen K (2004) Didactical structures as an outcome of research on teaching-learning sequences. Int J Sci Educ 26(5):537–554

Li Y, Wang K, Xiao Y, Froyd JE (2020) Research and trends in STEM education: a systematic review of journal publications. Int J STEM Educ 7(11)

Magnusson S, Krajcik J, Borko H (1999) Nature, sources, and development of PCK. In: Gess-Newsome J, Lederman NG (eds) PCK and science education. Kluwer Academic Publishers, pp 95–132

Martins I, Baptista M (2024) Teacher professional development in integrated STEAM education: a study on its contribution to the development of the PCK of physics teachers. Educ Sci 14(2):164

Marzari A, Di Mauro M, Rosi T, Onorato P, Malgieri M (2023) Investigating the principle of relativity and the principle of equivalence in classical mechanics: design and evaluation of a teaching-learning sequence based on experiments and simulations. Educ Sci 13:712

Masci C, De Witte K, Agasisti T (2018) The influence of school size, principal characteristics and school management practices on educational performance: an efficiency analysis of Italian students attending middle schools. Socio-Econ Plan Sci 61:52–69

McDermont LC (1990) Millikan Lecture 1990: What we teach and what is learned—Closing the gap. Am J Phys 59:301

McDermott LC, Shaffer PS, Constantinou CP (2000) Preparing teachers to teach physics and physical science by inquiry. Phys Educ 35:411–416

McDermott LC, Heron PRL, Shaffer PS, Stetzer MR (2006) Improving the preparation of K-12 teachers through physics education research. Am J Phys 74(9):763–767

Méheut M, Psillos D (2004) Teaching-learning sequences: aims and tools for science education research. Int J Sci Educ 26(5):515–535

Michelini M, Stefanel A (2014a) Prospective primary teachers and physics pedagogical content knowledge. In: Kaminski W, Michelini M (eds) Teaching and learning physics today: challenges? Benefits? Selected paper books of the international conference GIREP-ICPE-MPTL 2010, Reims 22–27 August 2010. Lithostampa, Udine, pp 149–157

Michelini M, Sperandeo Mineo RM (2014b) Challenges in primary and secondary science teachers education and training. In: Kaminski W, Michelini M (eds) Teaching and learning physics today. Lithostampa, Udine, pp 143–148

Michelini M, Santi L, Stefanel A, Vercellati S (2014) Community of prospective primary teachers facing the relative motion and PCK analysis. In: Kaminski W, Michelini M (eds) Teaching and learning physics today: challenges? Benefits? Proceedings of the international conference on GIREPICPE-MPTL 2010, Reims 22–27 August 2010. Lithostampa, Udine

Michelini M, Stefanel A (2015) Research based activities in teacher professional development on optics. Il Nuovo Cimento C 38:105–126

Mientus L, Hume A, Wulff P, Meiners A, Borowski A (2022) Modelling STEM teachers' pedagogical content knowledge in the framework of the refined consensus model: a systematic literature review. Educ Sci 12:385

Mincu M (2015) The Italian middle school in a deregulation era: modernity through path-dependency and global models. Comp Ed 51(3):446–462

Mincu C (2019) Learning to teach in Italy: reviewing policy and research trends. In: Tatto MT, Menter I (eds) Knowledge, policy and practice in teacher education, pp 135–152

Minner D, Levy AJ, Century J (2010) Inquiry-based science instruction—What is it and does it matter? Results from a research synthesis years 1984–2002. J Res Sci Teach 47:474

Mullins IVS, Martin MO (2008) TIMSS 2007. BostonCollege, Encyclopedia. Chestnut Hill

National Research Council (1996) From analysis to action: undergraduate education in science, mathematics, engineering, and technology. National Academy Press, Washington, D.C

National Research Council (2000) How people learn: brain, mind, experience, and school, expanded. National Academy Press, Washington, D.C

Nguyen TT, Duong NTT (2021) Designing experiential learning activities for students in directions of STEM education at primary schools. J Phys: Conf Ser 1835, 012056 (2021)

Ohlsson S (1996) Learning to do and learning to understand: a lesson and a challenge for cognitive modeling. In: Learning in humans and machines towards an interdisciplinary learning science, pp 37–62 (1996)
Olsen RV, Prenzel M, Martin R (2011) Interest in science: a many-faceted picture painted by data from the OECD PISA study. Int J Sci Educ 33(1):1–6
Park S, Oliver SJ (2008) Revisiting the conceptualisation of pedagogical content knowledge (PCK): PCK as a conceptual tool to understand teachers as professionals. Res Sci Educ 38:261–284
Santagata R, Yeh C (2016) The role of perception, interpretation, and decision making in the development of beginning teachers' competence. ZDM Math Educ 48:153–165
Schiering D, Sorge S, Keller MM, Neumann K (2022) A proficiency model for pre-service physics teachers' pedagogical content knowledge (PCK)—What constitutes high-level PCK? J Res Sci Teach 60(1):136–163
Schiering D, Sorge S, Keller MM, Neumann K (2023) A proficiency model for pre-service physics teachers' pedagogical content knowledge(PCK)—What constitutes high-level PCK? J Res Sci Teach 60(1):136–163
Shulman LS (1986) Those who understand: knowledge growth in teaching. Ed Res 15(Z):4–14
Shulman LS (1987) Knowledge and teaching: foundations of the new reform. Ed Res 57(1):1–22
Steffe LP, Thompson PW (2000) Teaching experiment methodology: underlying principles and essential elements. In: Lesh R, Kelly AE (eds) Handbook of research design in mathematics and science education. Erlbaum, Hillsdale, NJ, pp 267–307
Theodorakakos A, Psillos D (2010) PEC task explore. In: Constantinou C (ed) CBLIS2010 Warsaw: Oelizk, pp 75–83
Thornton RK, Sokoloff DR (1999) Learning motion concepts using RTL. Am J Phys 58(9):858–867
Tiberghien A, Buty C, Le Maréchal J-F (2005) Physics teaching sequences and students' learning. In: Science and technology education at cross roads: meeting the challenges of the 21st century. The second conference of EDIFE and the second IOSTE symposium in Southern Europe
Tuveri M, Fadda D, Steri A, Stefanizzi R, Gabriele F, Vivanet G, Bonivento W, Carbonaro CM, Fanti V (2023) Promoting the learning of modern and contemporary physics in high schools in informal and non-formal contexts. Il Nuovo Cimento C 46(6):12779
Viennot L (2008) Attracting students towards physics—A question of topics? Front Phys Educ 34–43
Vogelsang C, Kulgemeyer C, Riese J (2022) Learning to plan by learning to reflect?—Exploring relations between professional knowledge, reflection skills, and planning skills of preservice physics teachers in a one-semester field experience. Educ Sci 12:479
Vokos S, Seeley L, Etkina E (2022) Research-guided physics teaching: foundations, enactment, and outcomes. In: Guisasola J, McLoughlin E (eds) Connecting research in physics education with teacher education, vol 3. I.C.P.E. Book
White R, Gunstone R (1992) Probing understanding. The Falmer Press, London and New York

The manufacturer's authorised representative in the EU is Springer Nature Customer Service Centre GmbH, Europaplatz 3, 69115 Heidelberg, Germany. If you have any concerns regarding our products, please contact ProductSafety@springernature.com

Printed and bound by CPI Group (UK) Ltd, Croydon, CR0 4YY
30/09/2026
02225588-0006